Weston M. Stacey

Fusion

Related Titles

Stacey, W. M.

Nuclear Reactor Physics

2007
ISBN: 978-3-527-40679-1

Stacey, W. M.

Fusion Plasma Physics

2005
ISBN: 978-3-527-40586-2

Smirnov, B. M.

Plasma Processes and Plasma Kinetics

580 Worked-Out Problems for Science and Technology

2007
ISBN: 978-3-527-40681-4

Woods, L. C.

Theory of Tokamak Transport

New Aspects for Nuclear Fusion Reactor Design

2006
ISBN: 978-3-527-40625-8

Woods, L. C.

Physics of Plasmas

2004
ISBN: 978-3-527-40461-2

Diver, D.

A Plasma Formulary for Physics, Technology, and Astrophysics

2001
ISBN: 978-3-527-40294-6

Weston M. Stacey

Fusion

An Introduction to the Physics and Technology of
Magnetic Confinement Fusion

Second, Completely Revised and Enlarged edition

WILEY-VCH Verlag GmbH & Co. KGaA

The Author

Weston M. Stacey
Georgia Inst. of Technology
Nuclear & Radiological Engineering
and Medical Physics
900 Atlantic Drive, NW
Atlanta, GA 30332-0425
USA

Cover

With kind permission of
the ITER Organization,
St. Paul-les-Durance Cedex,
France.

1st Reprint 2011

Library of Congress Card No.: applied for

British Library Cataloguing-in-Publication Data
A catalogue record for this book is available from the
British Library.

**Bibliographic information published by
the Deutsche Nationalbibliothek**
The Deutsche Nationalbibliothek lists this
publication in the Deutsche Nationalbibliografie;
detailed bibliographic data are available on the
Internet at http://dnb.d-nb.de.

Typesetting Thomson Digital, Noida, India
Printing and Binding betz-druck GmbH, Darmstadt
Cover Design Adam-Design, Weinheim

Printed in the Federal Republic of Germany
Printed on acid-free paper

ISBN: 978-3-527-40967-9

Contents

Fusion: Second, Completely Revised and Enlarged edition. Weston M. Stacey
Copyright © 2010 Wiley-VCH Verlag GmbH & Co. KGaA, Weinheim
ISBN: 978-3-527-40967-9

Preface to the First Edition

The purpose of this book is to provide the reader with a broad and comprehensive introduction to the multidisciplinary subject of magnetic fusion. The material is self-contained in that each topic is developed from first principles to engineering application. Relevant physical property data are included where appropriate. The text evolved from lecture notes for a two-course sequence that is offered to advanced undergraduates and first-year graduate students at Georgia Tech.

An overview of the principles, status, and prospects for magnetic fusion is given in the first chapter. Basic properties of magnetically confined plasmas are developed and applied to establish the principles of the major confinement concepts in Chapters 2–4. The following two chapters deal with subjects involving in equal measure plasma physics and technology – plasma heating and the plasma–wall interaction. Major technological systems – magnets, energy storage and transfer, blanket and shield, tritium and vacuum – and their underlying technologies are treated in Chapters 7–11. The final chapter deals with the synthesis of the preceding material in fusion reactor design, emphasizing the interaction among the various physics and technology constraints. A summary of recent conceptual designs is provided.

The book is intended as a text for introductory courses on magnetic fusion physics and technology at the advanced undergraduate or first-year graduate level. It should also serve the needs of practicing engineers and scientists who wish to obtain a broad introduction to the subject. The material is accessible to anyone with a background equivalent to that of an advanced undergraduate in engineering or physics.

The quest after fusion power is exciting and demanding. The scientific, technical, and institutional challenges that must be met will require the dedication of the best minds that our society can produce for many years to come. But the ultimate reward for success will be enormous.

Atlanta, Georgia
January 1984

Weston M. Stacey, Jr.

Fusion: Second, Completely Revised and Enlarged edition. Weston M. Stacey
Copyright © 2010 Wiley-VCH Verlag GmbH & Co. KGaA, Weinheim
ISBN: 978-3-527-40967-9

Preface

A lot has happened in the quest for fusion power since the preface for the first edition of this book was written 25 years ago. The achieved value of the plasma confinement parameter has increased more than 10-fold, the radiation damage limits of structural materials have been increased 20-fold, large superconducting magnets have been operated at fields well in excess of 10 T, plasmas have been heated to thermonuclear temperatures with negative ion neutral beams and electromagnetic wave systems that did not exist at that time, deuterium-tritium plasmas have produced more than 10 MW of fusion power, and the first experimental power reactor (ITER) is under construction by an international team representing over half of the world's population. However, the quest for fusion power remains "exciting and demanding" and still "will require the dedication of the best minds that our society can produce for many years to come".

The purpose of this second edition remains the same – to provide a broad and comprehensive introduction to the multidisciplinary subject of magnetic confinement fusion. While the fundamentals remain the same, the treatment of the various plasma physics and fusion technology topics has been updated to reflect advances over the past quarter century and to describe the current state of the art.

The book is still intended as a text for an introductory course on the physics and technology of magnetic confinement fusion at the advanced undergraduate or first-year graduate level and as a self-study guide for practicing scientists and engineers who are entering the field. The material should be accessible to anyone with a background equivalent to that of an advanced undergraduate in physics or engineering.

Atlanta, Georgia
May, 2009

Weston M. Stacey

Fusion: Second, Completely Revised and Enlarged edition. Weston M. Stacey
Copyright © 2010 Wiley-VCH Verlag GmbH & Co. KGaA, Weinheim
ISBN: 978-3-527-40967-9

Acknowledgments to the First Edition

The material for this book was collected from many sources too numerous to mention here. Some of the published sources are cited throughout the text.

Several of my colleagues were generous enough with their time to review parts of the text and to offer helpful suggestions: D.B. Montgomery and R.J. Thome – the magnetics chapter; R.L. Kustom – the energy storage and transfer chapter; V.A. Maroni and P.A. Finn – the tritium and vacuum chapter. The manuscript was prepared by Mrs. Barbara Cranfill. To all of these people I express my gratitude.

W.M.S.

Acknowledgments

The new material included in this second edition was collected from many sources too numerous to cite in the literature of the field. However, the author wishes to express his appreciation for their expert assistance in sorting through some of this material to Mohamed Abdou of UCLA and Steve Zinkle of Oak Ridge National Laboratory who were good enough to provide summary material of recent advances in tritium breeding blankets and materials, respectively. The author is also grateful to his administrative assistants, Valerie Spradling and Shauna Bennett-Boyd, for their assistance over recent years in assembling the new material into a manuscript, and to the students in the 2009 Introduction to Fusion class for help with the proofreading. Finally, no book can be brought to life without the efforts of the people that publish it, and the author is grateful to Anja Tschörtner and others at Wiley-VCH for suggesting and producing this second edition.

W.M.S.

Fusion: Second, Completely Revised and Enlarged edition. Weston M. Stacey
Copyright © 2010 Wiley-VCH Verlag GmbH & Co. KGaA, Weinheim
ISBN: 978-3-527-40967-9

1
Introduction

1.1
Basics of Fusion

Fusion has the potential of providing an essentially inexhaustible source of energy for the future. Under the proper conditions the low atomic number elements will react to convert mass to energy $[E = (\Delta m)c^2]$ via nuclear fusion. For example, the fusion of the hydrogen isotopes deuterium (D) and tritium (T) according to the reaction

$$D + T \rightarrow {}^4He + n \quad (Q = 17.6 \, MeV)$$

produces 17.6 MeV of energy.

The fusion of 1 g of tritium (together with $\frac{2}{3}$ g of deuterium) produces 1.6×10^5 kW-hr of thermal energy. Deuterium exists at 0.0153 at.% in sea-water and is readily extractable, thus constituting an essentially infinite fuel source. Tritium undergoes beta decay with a half-life of 12.5 yr. Thus it must be produced artificially, for example, by neutron capture in lithium. (Natural lithium consists of 7.5% ${}^6Li$ and 92.5% ${}^7Li$ and is quite abundant in nature). The tritium production reactions are

$$n + {}^6Li \rightarrow T + {}^4He$$
$$n + {}^7Li \rightarrow T + {}^4He + n'$$

The first reaction has a large cross section for thermal (slow) neutrons, while the second reaction is more probable with fast neutrons. When the lithium is placed around the fusion chamber, the fusion neutrons can be used to produce tritium, thereby introducing the possibility of a fusion reactor "breeding" its own fuel.

Another promising fusion reaction is

$$D + D \rightarrow \begin{cases} T + H & (Q = 4 \, MeV) \\ {}^3He + n & (Q = 3.25 \, MeV) \end{cases}$$

which has two branches of roughly equal probability. With this reaction the fusion fuel source is truly inexhaustible.

A third possible fusion reaction is

$$D + {}^3He \rightarrow {}^4He + H \quad (Q = 18.2 \, MeV)$$

There are many others involving the low atomic number elements.

Fusion: Second, Completely Revised and Enlarged edition. Weston M. Stacey
Copyright © 2010 WILEY-VCH Verlag GmbH & Co. KGaA, Weinheim
ISBN: 978-3-527-40967-9

In order for a fusion reaction to take place the two nuclei must have enough energy to overcome the repulsive Coulomb force acting between the nuclei and approach each other sufficiently close that the short-range attractive nuclear force becomes dominant. Thus, the fusion fuel must be heated to high temperatures. For the D-T reaction, the gas temperature must exceed 5×10^7 K before a significant fusion rate is feasible. An even higher gas temperature is required for the other fusion reactions. At such temperatures the gas exists as a macroscopically neutral collection of ions and unbound electrons which is called a plasma.

The fusion reaction rate per unit volume can be written

$$R(\text{fusions}/\text{m}^3) = n_1 n_2 \langle \sigma v \rangle_{12}$$

where n_1 and n_2 are the densities of species 1 and 2, respectively, and

$$\langle \sigma v \rangle_{12} = \iint d^3 v_1 d^3 v_2 f_1(\mathbf{v}_1) f_2(\mathbf{v}_2) |\mathbf{v}_1 - \mathbf{v}_2| \sigma_f(|\mathbf{v}_1 - \mathbf{v}_2|)$$

is the fusion reactivity. Here v is the velocity, f is the velocity distribution function, and σ_f is the fusion cross section. It is usually adequate to use a Maxwellian distribution,

$$f_{\text{max}} = \left(\frac{m}{2\pi kT} \right)^{3/2} e^{-mv^2/2kT}$$

to evaluate $\langle \sigma v \rangle$, in which case the value of the integral depends only upon the temperature T of the plasma. The fusion reactivity is shown in Figure 1.1 for the three reactions cited above. The D-T fusion reactivity is much greater than that for other

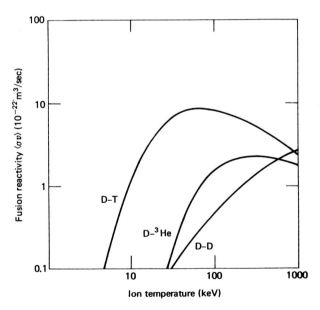

Figure 1.1 Fusion reactivity.

potential fusion reactants, which is the reason why achieving the necessary conditions for D-T fusion is the principal goal of the present phase of fusion research.

The 17.6 MeV of energy produced in the D-T fusion event is in the form of kinetic energy of the neutron (14.1 MeV) and the alpha particle (3.5 MeV). The alpha particle is confined within the plasma, and its energy is distributed via collisions among the plasma ions and electrons and eventually is incident upon the wall of the reaction chamber as a surface heat flux. The neutron leaves the plasma immediately and gives up its energy to the surrounding material via elastic and inelastic collisions with lattice atoms, producing a volume heat source and, in the process, radiation damage to the material due to atomic displacements. The fusion neutron is ultimately captured or leaks from the system. The neutron may be captured in lithium to produce tritium to sustain the fusion fuel cycle, or it may be captured in structural material, thereby producing a relatively short-lived radioactive isotope. In projected applications of fusion to produce fuel for fission reactors, the neutron may be captured in uranium-238 or thorium-232, thereby initiating nuclear transmutation chains leading ultimately to plutonium-239 or uranium-233, respectively.

Heating a plasma to thermonuclear temperatures and then confining it sufficiently well that a net positive energy balance can be achieved are the two premier issues which will determine the scientific feasibility of fusion. Experimental progress with magnetically confined plasmas has been impressive in recent years, and scientific feasibility should be convincingly established in the near future.

Beyond scientific feasibility there are several scientific and technological issues which will determine the practicality of fusion. Among the former are achieving high power density and a quasi-steady-state mode of operation. Technological issues include plasma heating systems, superconducting magnets, materials, tritium breeding blankets, vacuum systems, and remote maintenance.

1.2
Magnetic Confinement

A fusion plasma cannot be maintained at thermonuclear temperatures if it is allowed to come in contact with the walls of the confinement chamber, because material eroded from the walls would quickly cool the plasma. Fortunately, magnetic fields can be used to confine a plasma within a chamber without contact with the wall. (This statement is only approximately true, as we shall see in a later chapter).

A charged particle moving in a magnetic field will experience a Lorentz force which is perpendicular to both the direction of particle motion and to the magnetic field direction. This force does not affect the component of particle motion in the magnetic field direction, but it causes acceleration at right angles to the particle direction in the plane perpendicular to the magnetic field direction, producing a circular particle motion in that plane. Thus, a particle in a magnetic field will move along the field and circle about it; that is, will spiral about the field line. The radius of the spiral, or gyroradius, is inversely proportional to the strength of the magnetic field, so that in a strong field charged particles move along magnetic field lines, as shown in Figure 1.2.

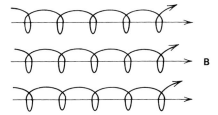

Figure 1.2 Motion of charged particles along magnetic field lines.

1.2.1
Closed Toroidal Confinement Systems

The magnetic field lines may be configured to remain completely within a confinement chamber by the proper choice of position and currents in a set of magnetic coils. The simplest such configuration is the torus, shown in Figure 1.3. A set of coils can be placed to produce a toroidal field B_ϕ. Particles following along the closed toroidal field lines would remain within the toroidal confinement chamber.

The curvature and nonuniformity of the toroidal field produce forces which act upon the charged particles to produce "drift" motions that are radially outward, which would, if uncompensated, cause the particles to hit the wall. A poloidal magnetic field must be superimposed upon the toroidal magnetic field in order to compensate these drifts, resulting in a helical magnetic field which is entirely contained within the toroidal confinement chamber. This poloidal field may be produced by a toroidal current flowing in the plasma (tokamak) or by external coils (stellarator, etc.).

The tokamak concept, which was invented in the U.S.S.R. in the mid-1960s, has been the most extensively investigated worldwide and is the most advanced. This

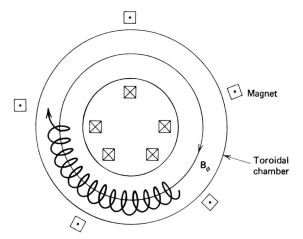

Figure 1.3 Closed toroidal confinement.

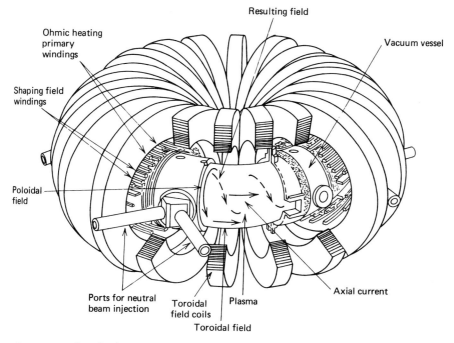

Ohmic heating
primary
windings

Resulting field

Vacuum vessel

Shaping field
windings

Poloidal
field

Ports for neutral
beam injection

Toroidal
field coils

Plasma

Toroidal field

Axial current

Figure 1.4 Tokamak schematic.

concept is illustrated in Figure 1.4. The toroidal field is produced by a set of toroidal field coils which encircle the plasma. The poloidal field is produced by an axial, or toroidal, current which is induced by the transformer action of a set of primary poloidal field, or ohmic heating, coils. The extensive worldwide interest in the tokamak concept is indicated by the partial lists of experiments given in Table 1.1. The relative scale of some of these devices is illustrated in Figure 1.5.

Table 1.1 Representative tokamak experiments.

Device	Major Radius (m)	Minor Radius (m)	Magnetic Field (T)	Plasma Current (MA)
T-10 (Russia)	1.5	0.37	4.5	0.7
TFTR (USA)	2.4	0.80	5.0	2.2
JET (UK)	3.0	1.25	3.5	7.0
DIII-D (USA)	1.7	0.67	2.1	2.1
ToreSupra (FR)	2.4	0.80	4.5	2.0
ASDEX-U (Ger)	1.7	0.50	3.9	1.4
JT60-U (Japan)	3.4	1.1	4.2	5.0
ALCMOD (USA)	0.7	0.22	9.0	1.1
EAST (China)	1.7	0.4	3.5	1.0
K-STAR (Korea)	1.8	0.5	3.5	2.0

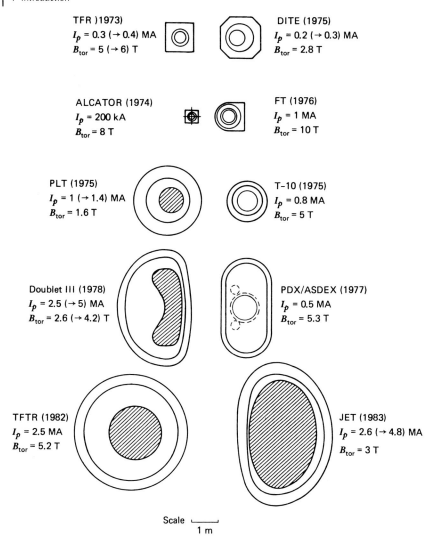

Figure 1.5 Relative size of some major tokamak devices throughout the world. Dimensions are to scale and the plasma current and the magnetic field are given in each case.

1.2.2
Open (Mirror) Confinement Systems

It is possible to confine a plasma magnetically within a fixed confinement chamber, even when the magnetic field lines themselves do not remain within the confinement chamber, by trapping the charged particles in a magnetic well. The principle is illustrated in Figure 1.6, where the particle speeds along $(v_{\parallel})$ and perpendicular to $(v_{\perp})$ the magnetic field are denoted.

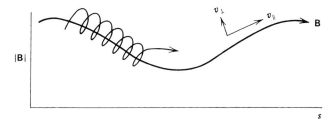

Figure 1.6 A magnetic well.

Because the force exerted on a moving charge by a magnetic field is orthogonal to the direction of particle motion, no work is done on the particle and its kinetic energy (KE) remains constant:

$$\frac{1}{2} m \left[v_{\parallel}^2(s) + v_{\perp}^2(s) \right] \equiv KE = \text{constant}$$

The angular momentum is also conserved, so that

$$\frac{1}{2} \frac{m v_{\perp}^2(s)}{B(s)} \equiv \mu = \text{constant}$$

Thus, the velocity along the field line can be expressed as

$$v_{\parallel}^2(s) = \frac{2}{m} \left[KE - \mu B(s) \right]$$

For particles with sufficiently large values of μ, the right-hand side (RHS) can vanish as the particle moves along the field line into a region of stronger magnetic field. When the field is strong enough that the RHS vanishes, the particle is reflected and travels back along the field line through a region of reducing, then increasing magnetic field strength until it comes once again to a reflection point, at which the field strength is again large enough to make the RHS vanish. This is the principle of the simple mirror illustrated in Figure 1.7, in which a magnetic well of the type shown in Figure 1.6 is created by ring coils carrying currents of different magnitudes.

The simple mirror is unstable against flute-type instabilities in which the plasma bulges outward in the low field regions. The flute-type instability can be suppressed by placing the plasma in a three-dimensional magnetic well. Such a "minimum-B" magnetic configuration can be created by a coil wound like the seams of a baseball, as shown in Figure 1.7. The minimum-B mirror concept is well understood based on the results of more than a dozen experiments, but the prospects for achieving a favorable power balance in a fusion reactor are not good.

Because electrons escape faster than ions, mirrors operate with a slightly positive electrostatic potential. This fact allows minimum-B mirrors to be used at each end of a simple central cell mirror to confine ions electrostatically in the central cell in the tandem mirror concept. The present mirror experiments in the world are indicated in Table 1.2.

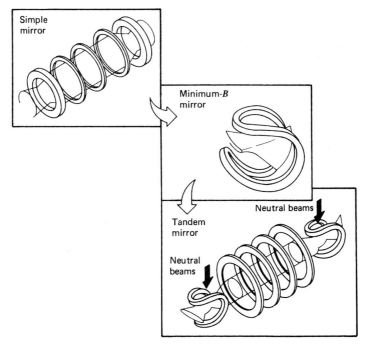

Figure 1.7 The evolution of magnetic mirrors. (Reproduced by permission of university of California, Lawrence Livermore National Laboratory, and the Department of Energy.)

Table 1.2 Mirror experiments.

Device	Country
GAMMA-10	Japan
GOL-3	Russia
GDT	Russia
AMBAL-M	Russia
HANBIT	Korea

1.3
Feasibility of Fusion

Feasibility issues can be conveniently separated into issues of scientific, engineering and economic feasibility, at least for the purpose of discussion.

1.3.1
Scientific Feasibility

Scientific feasibility basically comes down to confining a plasma at thermonuclear temperatures well enough that a sufficiently positive energy balance can be obtained to enable net power to be produced from the fusion reactions. The plasma power amplification factor

$$Q_p = \frac{\text{fusion power}}{\text{external heating power}}$$

is a conventional measure of scientific feasibility, albeit on a sliding scale. $Q_p > 1$ will constitute "breakeven" on the plasma energy balance when achieved (i.e., the plasma will be producing a larger amount of thermal power from fusion reactions than the amount of external power that must be provided in addition to the fusion power to maintain the plasma at thermonuclear temperature). The ultimate goal (actually the holy grail) of plasma physics research is the achievement of sufficiently good confinement that the 20% of the D-T fusion energy retained in the plasma in the form of an energetic alpha particle is sufficient to maintain the plasma at thermonuclear temperature without any external power; this condition of $Q_p = \infty$ is known as "ignition". However, no matter how good the confinement, future fusion reactors will operate with some external power for control purposes, so the practical definition of scientific feasibility is Q_p large enough that net electrical power can be economically produced, which is probably $Q_p > 10$.

The plasma power balance can be written as

$$\textit{fusion heating} + \textit{external heating} \geq \textit{radiation loss} + \textit{transport loss}$$

$$\frac{1}{4} n^2 \langle \sigma v \rangle_{fus} U_\alpha \left(1 + \frac{5}{Q_p} \right) \geq f_z n^2 L_z + \frac{3nT}{\tau_E}$$

where $n = n_D + n_T$ is the total ion density, $U_\alpha = 3.5$ MeV is the energy of the alpha particle from D-T fusion, f_z and L_z are the concentration and radiation emissivity of the ions (plasma plus impurity) present, T is the plasma temperature, τ_E is the time which energy is confined in the plasma before escaping, and $\langle \sigma v \rangle_{fus} \simeq \text{const} \times T^2$ is the fusion reactivity. Neglecting the radiation term, this equation can be rearranged to obtain the Lawson criterion for ignition

$$nT\tau_E \geq \frac{12}{\text{const} \times U_\alpha} \equiv (nT\tau_E)_{\text{Lawson}}$$

The achieved values of the plasma energy amplification factor Q_p and of the triple product $nT\tau_E$ are indicated in Figure 1.8.

The triple product $nT\tau_E = p\tau_E$, indicating that achievement of power balance at high Q_p requires achieving high plasma pressure and long plasma energy confinement time. The magnitude of plasma pressure that can be achieved is related to the magnitude of the confining magnetic field pressure magnetohydrodynamic (MHD) instability limits which can be characterized in terms of limiting values of

$$\beta \equiv \frac{\textit{plasma pressure}}{\textit{magnetic pressure}} = \frac{nkT}{B^2/2\mu_0}$$

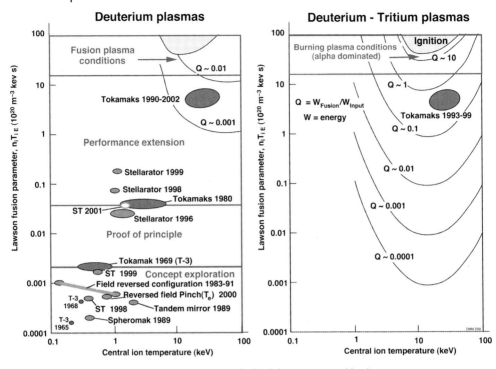

Figure 1.8 Progress towards scientific feasibility as measured by the Lawson criterion.

Since the power density in a D-T plasma can be written

$$P = \frac{1}{4}n^2 \langle \sigma v \rangle_{fus} U_{fus} \simeq \frac{1}{4}\text{const} \times (nT)^2 U_{fus} \sim \beta^2 B^4$$

a second aspect of practical scientific feasibility is the achievement of sufficiently large values of β.

1.3.2
Engineering Feasibility

The development of magnetic fusion as an energy source requires the development of two types of engineering technologies – (i) those "plasma support" technologies that are necessary to heat and confine the plasma, and (ii) those "nuclear" technologies that are necessary for a power reactor.

The plasma support technologies include the vacuum system, the magnet system, the plasma heating system, the plasma fueling system and other such systems. The development of plasma-facing components that can survive in the high particle and heat fluxes that will be incident upon them in ITER and future fusion reactors is also included. These systems are relatively highly developed as a result of magnetic fusion R&D that has been carried out world-wide in support of national fusion programs for the past 40 years and more recently in the R&D program that has been carried out for the International Thermonuclear Experimental Reactor (ITER). The operating parameters of systems that are being developed for ITER are comparable to those that will be needed for future power reactors, and ITER operation will provide a test of these technologies in a fusion reactor environment. While the requirements placed on these technologies by ITER and future reactors are certainly challenging, these challenges would appear to be more a matter of achieving reliable operation in a complex environment than of achieving a breakthrough in performance capability.

The state of development is less advanced for those "nuclear" technologies needed, in addition to reliable "plasma support" technologies, for a fusion power reactor. The technology needed to "breed" the tritium required for D-T plasma operation is being developed for testing in ITER, and the technology needed to process tritium is being developed for the operation of ITER. The heat removal and power conversion technologies needed for fusion power are similar to those needed for conventional (fission) nuclear power. The development of radiation-resistant structural materials that can be used in the intense fast neutron environment that will be present in a fusion reactor is a major challenge.

1.3.3
Practical Feasibility

The ultimate goal of fusion development – a power plant that provides power to an electrical power grid reliably and with high availability – creates some practical feasibility issues for fusion plasma physics and technology related to sustainable modes of operation. For example, since continuous or quasi-continuous operation is preferable to short-pulse operation with long down-times between pulses, the development of non-inductive current-drive techniques may be a feasibility issue for the conventional tokamak confinement concept, which is inherently pulsed. On the other hand, the requirement for very small tolerances in the manufacture of the large and very complex magnetic coils of a stellarator (which is inherently steady-state) constitutes another type of practical feasibility issue.

1.3.4
Economic Feasibility and Fuel Resources

It is possible to estimate the cost of building and operating a fusion power plant to produce electricity in the middle of the century and to compare with a similar estimate of the cost of building and operating a coal-fired or nuclear plant then, and this has been done. The cost estimate of fusion electricity is usually somewhat more expensive, using today's cost formulas. However, today's cost formulas do not include things like the cost of preventing carbon emissions into the atmosphere, or the costs of recovering more than the 1% of the energy potential of uranium that is recovered in the present "once-through" nuclear fuel cycle, or the availability into the future of the fuel resource. If the world's proven reserves of fossil fuels, uranium and lithium (source of tritium) were each "burned" at the rate required to provide all of the world's estimated electricity usage in 2050, the fossil fuels would be gone in less than 100 years, the uranium used in the present "once-through" fuel cycle would be gone by the end of the century (this could be extended to a few hundred years by the introduction of fast breeder reactors), but the lithium would last more than 6000 years.

2
Basic Properties

2.1
Plasma

A plasma is a macroscopically neutral collection of ions and electrons. On a microscopic scale, of course, the plasma is nonuniform and nonneutral. If n_0 is the average ion or electron density and $n_e(r)$ is the actual local electron density, there exists an electrostatic potential $\phi(r)$ which satisfies Poisson's equation

$$\nabla^2 \phi(r) = \frac{e(n_e - n_0)}{\varepsilon_0}$$

This electrostatic potential determines the local electron density, which latter is given by the Maxwell–Boltzmann distribution,

$$n_e(r) = n_0 \exp\left(\frac{e\phi}{kT_e}\right) \simeq n_0\left(1 + \frac{e\phi}{kT_e}\right)$$

Combining these results yields

$$\nabla^2 \phi(r) = \frac{n_0 e^2}{\varepsilon_0 kT_e}\phi(r) \equiv \frac{1}{\lambda_e^2}\phi(r)$$

which has the solution

$$\phi(r) = \left(\frac{e}{4\pi\varepsilon_0 r}\right)\exp\left(\frac{-r}{\lambda_e}\right)$$

In the limit $r \to 0$, this solution is just the potential due to an isolated point charge (the term in the first brackets). The exponential term represents an attenuation of the potential, or shielding, due to other charged particles. The *e*-folding distance of the shielded potential,

$$\lambda_e = \left(\frac{\varepsilon_0 kT_e}{n_0 e^2}\right)^{1/2}$$

is known as the Debye length.

Fusion: Second, Completely Revised and Enlarged edition. Weston M. Stacey
Copyright © 2010 WILEY-VCH Verlag GmbH & Co. KGaA, Weinheim
ISBN: 978-3-527-40967-9

The validity of this type of analysis, which deals with a distribution of electrons rather than isolated charges and which leads to a shielded potential, requires that the number of electrons within $r < \lambda_e$ be large; that is,

$$n_\lambda \equiv \frac{4}{3} \pi \lambda_e^3 n_0 = \frac{4}{3} \pi \left(\frac{\varepsilon_0 k T_e}{n_0 e^2} \right)^{3/2} n_0 \gg 1$$

This condition, together with the requirement that the physical dimension of the system L be large compared to a Debye length – $L \gg \lambda_e$ – defines a plasma. (A similar argument can be outlined for the ions, leading to an analogous definition of the ion Debye length). For a fusion plasma, $\lambda \simeq 10^{-4} - 10^{-5}$ m and $n_\lambda \simeq 10^6 - 10^{10}$.

In principle, the properties of a plasma can be determined by solving simultaneously the equations of motion for all particles in the plasma acting under forces due to the electrostatic interaction with all the other particles and due to any external magnetic and electric fields. Since there are typically approximately 10^{20} particles per cubic meter in plasmas of interest for fusion, this approach is impractical.

The fact that the electrostatic potential due to a point charge is effectively shielded within a Debye length due to the presence of other charges suggests a collective treatment of the force exerted on a particle by all the other particles. Thus, the usual procedure involves the calculation of the electric and magnetic fields due to all the particles and then the calculation of the motion of the plasma particles under the forces arising from these internal fields as well as external fields.

This treatment is valid for interactions among particles which take place at distances greater than the Debye length. However, a small number of interactions take place at distances shorter than a Debye length. These latter, close-range interactions are treated as two-body "scattering" events.

Thus, brief reviews of electric and magnetic fields and of Coulomb scattering are in order.

2.2
Electric and Magnetic Fields

Gauss's law states that the integral of the electric displacement vector **D** over any closed surface S is equal to the volume integral of the charge density ρ over the volume V contained within the surface,

$$\int_S \mathbf{D} \cdot d\mathbf{s} = \int_V \rho \, d^3 r$$

Using the divergence theorem,

$$\int_S \mathbf{D} \cdot d\mathbf{s} \equiv \int_V \nabla \cdot \mathbf{D} \, d^3 r$$

and the fact that the volume in question is arbitrary, leads to the differential form of Gauss' law,

$$\nabla \cdot \mathbf{D} = \rho$$

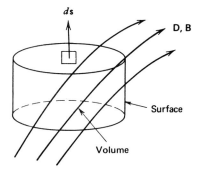

Figure 2.1 Arbitrary volume element.

Similarly, since there is no magnetic equivalent of charge, the solenoidal law of magnetic fields **B** is

$$\int_s \mathbf{B} \cdot d\mathbf{s} = 0$$

and using the divergence theorem yields the solenoidal law

$$\nabla \cdot \mathbf{B} = 0$$

See Figure 2.1.

Faraday's law states that a changing magnetic flux Φ through an arbitrary closed loop in space produces an electromotive force around that loop,

$$\oint \mathbf{E} \cdot d\mathbf{l} = -\frac{d\Phi}{dt} = \frac{-d}{dt} \int_s \mathbf{B} \cdot d\mathbf{s}$$

where **E** is the electric field and s is any closed surface bound by the loop. Using Stokes' theorem,

$$\oint \mathbf{E} \cdot d\mathbf{l} \equiv \int_s \nabla \times \mathbf{E} \cdot d\mathbf{s}$$

and the arbitrariness of the surface yields the differential form of Faraday's law.

$$\nabla \times \mathbf{E} = \frac{-\partial \mathbf{B}}{\partial t}$$

(assuming that the loop itself does not change in time).

Ampere's law states that the magnetomotive force around a closed loop is equal to the current flowing across an arbitrary surface bound by that loop plus the time rate of change of the electric displacement integrated over that surface,

$$\oint \mathbf{H} \cdot d\mathbf{l} = \int_s \mathbf{j} \cdot d\mathbf{s} + \frac{d}{dt} \int_s \mathbf{D} \cdot d\mathbf{s}$$

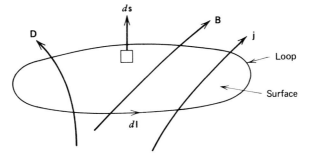

Figure 2.2 Arbitrary loop.

where **H** is the magnetic displacement and **j** is the current density. Use of Stokes' theorem leads to the differential form of Ampere's law,

$$\nabla \times \mathbf{H} = \mathbf{j} + \frac{\partial \mathbf{D}}{\partial t}$$

See Figure 2.2.

Poisson's equation describes the electrostatic potential ϕ associated with a charge ρ,

$$\nabla^2 \phi = -\frac{\rho}{\varepsilon_0}$$

where ε_0 is the permittivity. This equation has the solution

$$\phi(r) = \frac{1}{4\pi\varepsilon_0} \int_v \frac{\rho(r')}{|\mathbf{r} - \mathbf{r}'|} d^3 r'$$

The electric field and electric displacement are related by a constitutive relation,

$$\mathbf{D} = \varepsilon_0 \mathbf{E}$$

and similarly for the magnetic field and magnetic displacement,

$$\mathbf{B} = \mu_0 \mathbf{H}$$

where μ_0 is the permeability.

Comparing the vector identity

$$\nabla \cdot (\nabla \times \mathbf{A}) = 0$$

where **A** is any vector, with the solenoidal law of magnetic fields, $\nabla \cdot \mathbf{B} = 0$, it is seen that the magnetic field can be written as the curl of a vector potential **A**,

$$\mathbf{B} = \nabla \times \mathbf{A}$$

Using this form in Faraday's law leads to

$$\nabla \times \left(\mathbf{E} + \frac{\partial \mathbf{A}}{\partial t} \right) = 0$$

When this result is compared with the vector identity

$$\nabla \times \nabla \phi = 0$$

it follows that the electric field can be expressed as the gradient of an electrostatic potential plus the time derivative of the vector potential,

$$\mathbf{E} = -\nabla\phi - \frac{\partial \mathbf{A}}{\partial t}$$

Use of the vector potential representation of $\mathbf{B}$ in the time-independent version of Ampere's law leads to

$$\nabla^2 \mathbf{A} = -\mu_0 \mathbf{j}$$

which has the solution

$$\mathbf{A}(r) = \frac{\mu_0}{4\pi} \int_v \frac{\mathbf{j}(r')}{|\mathbf{r}-\mathbf{r}'|} d^3 r'$$

Thus, if the distribution and motion of the particles are known, ρ and $\mathbf{j}$ can be determined, which allows the electrostatic (ϕ) and vector ($\mathbf{A}$) potentials to be constructed. With these potentials, the electric and magnetic fields can then be computed.

2.3
Plasma Frequency

As an example of collective plasma phenomena, consider an initially uniform plasma slab in which all the electrons in the interval $x_1 < x < x_0$ are magically displaced to x_1 (Figure 2.3). Since the electrons are much less massive than the ions, their motion can be considered under the simplifying assumption that the ions remain fixed.

The net excess charge at x_1 is

$$\rho = -n_0 e(x_0 - x_1)$$

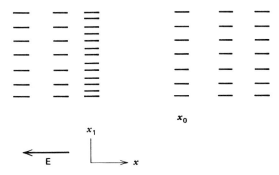

Figure 2.3 Slab plasma example.

where n_0 is the uniform, unperturbed density. From Poisson's equation, this charge produces an electric field

$$E(x_1) = -\nabla\phi = \frac{n_0 e}{\varepsilon_0}(x_0 - x_1)$$

in the negative x direction.

The equation of motion for the electrons at x_1, is

$$m_e \ddot{x}_1 = eE(x_1) = \frac{e^2 n_0}{\varepsilon_0}(x_0 - x_1)$$

Introducing $\xi \equiv x_0 - x_1$, this equation becomes

$$\ddot{\xi} + \left(\frac{e^2 n_0}{m_e \varepsilon_0}\right)\xi = 0$$

which has the oscillatory solution

$$\xi(t) = Ae^{i\omega_{pe}t} + Be^{-i\omega_{pe}t}$$

Thus, the electrons initially at x_1, are accelerated toward x_0 by the field, overshoot, with a reversal of the field, and oscillate back and forth about x_0 – much as would a spring which is initially stretched, then released. The oscillation frequency

$$\omega_{pe} = \left(\frac{e^2 n_0}{m_e \varepsilon_0}\right)^{1/2}$$

is known as the electron plasma frequency, and is on the order of 10^{10}–10^{11} rad/sec for plasmas of fusion interest.

An ion plasma frequency can be defined by analogy,

$$\omega_{pi} = \sqrt{\frac{m_e}{m_i}}\omega_{pe} \simeq \frac{1}{50}\omega_{pe}$$

2.4
Coulomb Scattering

Relatively short-range (within a Debye length) interactions are responsible for a number of important effects in plasmas. Even though these short-range interactions, which are referred to as scattering events, are much less numerous than the longer range interactions which can be treated collectively in terms of electric and magnetic fields, the strength of the short-range interactions is much stronger because of the nature of the electric force $F = -\nabla\phi \sim 1/r^2$.

Consider a charged particle 1 moving with velocity v_1 toward a stationary particle 2. Let x be the distance of closest approach if there was no force of interaction (Figure 2.4). If the two particles are charged, there is a force

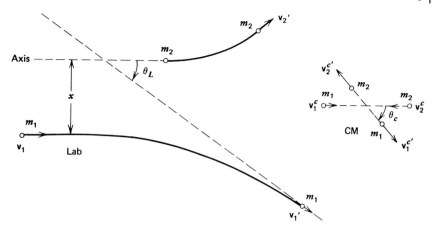

Figure 2.4 Scattering interaction.

$$F_{12} = \frac{e_1 e_2 (\mathbf{r}_1 - \mathbf{r}_2)}{4\pi\varepsilon_0 |\mathbf{r}_1 - \mathbf{r}_2|^3}$$

acting along the line connecting the two particles. This force is attractive if the particles have opposite charges, and repulsive if both have the same sign charge. Taking the latter case, the equations of motion are

$$F_{12} = m_1 \ddot{\mathbf{r}}_1 \quad -F_{12} = m_2 \ddot{\mathbf{r}}_2$$

Defining the separation $\mathbf{r} \equiv \mathbf{r}_1 - \mathbf{r}_2$, these equations may be added to obtain

$$-\frac{m_1 m_2}{m_1 + m_2} \ddot{\mathbf{r}} \equiv m_r \ddot{\mathbf{r}} = \frac{e_1 e_2 \mathbf{r}}{4\pi\varepsilon_0 r^3}$$

The center of mass (CM) is moving with a velocity $\mathbf{v}_c$ defined by

$$(m_1 + m_2)\mathbf{v}_c = m_1 \mathbf{v}_1$$

The scattering angle θ_c in the coordinate system that moves with the CM can be related to the scattering angle θ_L in the fixed (lab) coordinate system by making use of the conservation of kinetic energy and of momentum,

$$\cot\theta_L = \frac{m_1}{m_2} \csc\theta_c + \cot\theta_c$$

Some special cases are

$$m_1 \ll m_2 \quad \theta_L \simeq \theta_c$$

$$m_1 = m_2 \quad \theta_L = \frac{1}{2}\theta_c$$

$$m_1 \gg m_2 \quad \theta_L = \frac{m_2}{m_1}\sin\theta_c$$

The solution of the equation of motion is most readily carried out in the CM system, yielding a relation between the scattering angle and the impact parameter,

$$\tan\left(\frac{\theta_c}{2}\right) = \frac{|e_1 e_2|}{4\pi\varepsilon_0 m_r v^2 x}$$

where $v \equiv |\mathbf{v}_1 - \mathbf{v}_2|$ is the relative speed of approach of the two particles. From this result it is seen that the scattering angle increases as the impact parameter x decreases and as the relative speed decreases.

The probability that a particle will scatter through an angle of 90° or more in the CM system in a single interaction is just the probability that the collision has an impact parameter less than x_{90}, where

$$x_{90} = \frac{|e_1 e_2|}{4\pi\varepsilon_0 m_r v^2 \tan(\pi/4)}$$

Thus,

$$\sigma(\theta_c > 90°) = \pi x_{90}^2 = \frac{(e_1 e_2)^2}{\pi(4\varepsilon_0 m_r v^2)^2}$$

The probability that a particle will scatter through 90° or more as a result of multiple small-angle collisions is much larger. Consider a particle that travels a distance L in a gas of density n_2. The mean square deflection is

$$\overline{(\Delta\theta_c)^2} = n_2 L \int_{x_{min}}^{x_{max}} (\Delta\theta_c)^2 2\pi x dx$$

where, for small-angle scattering,

$$\Delta\theta_c \simeq 2\tan\left(\frac{\Delta\theta_c}{2}\right) = \frac{|e_1 e_2|}{2\pi\varepsilon_0 m_r v^2 x}$$

and we take $x_{max} = \lambda_2$ (the Debye length) and $x_{min} = x_{90}$ (certainly 90° is an upper limit on a "small" angle). Then

$$\overline{(\Delta\theta_c)^2} = \frac{n_2 L}{2\pi}\left(\frac{e_1 e_2}{\varepsilon_0 m_r v^2}\right)^2 \ln\Lambda$$

where

$$\Lambda \equiv 12\pi \left[\frac{(\varepsilon_0 kT)^3}{n_2 e_2^4 e_1^2}\right]^{1/2}$$

This expression can be used to determine the mean free path for 90° deflection by setting $\Delta\theta_c = \pi/2 \simeq 1$ and solving for L_{90}. The corresponding scattering probability is then

$$\sigma_{90}^{c} \equiv \frac{1}{n_2 L_{90}} = \frac{(e_1 e_2)^2 \ln \Lambda}{2\pi (\varepsilon_0 m_r v^2)^2}$$

Comparing,

$$\frac{\sigma_{90}^{c}}{\sigma(\theta_c > 90°)} = 8 \ln \Lambda \simeq 10^2$$

it is apparent that large-angle deflection is due almost entirely to multiple small-angle scattering events.

Scattering interactions are much more numerous than fusion interactions, as indicated in Figure 2.5. Thus, scattering interactions have a very important effect upon plasma properties.

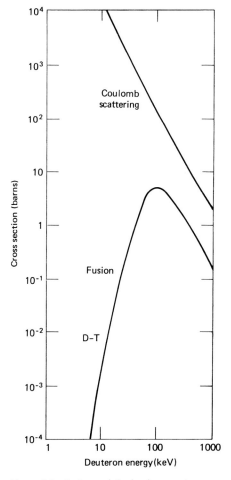

Figure 2.5 Fusion and Coulomb scattering cross sections.

2.5
Characteristic Times

The time required for a particle to scatter through $90°$ is a reasonable measure of the time required for that particle to lose any ordered momentum it may have originally had. The momentum transfer time for a particle of type 1 scattering from particles of type 2 is then

$$\left(\tau_{90}^{12}\right)_c \equiv \frac{L_{90}}{v_1} = \frac{2\pi\sqrt{m_r}\varepsilon_0^2(3kT)^{3/2}}{n_2(e_1e_2)^2\ln\Lambda}$$

where $\frac{1}{2}mv_1^2 = \frac{3}{2}kT$ has been used. The subscript c denotes that this is the time required to scatter $90°$ (via multiple small-angle interactions) in the CM system. Referring back to the relationship between θ_L and θ_c in the previous section, it can be seen that the time to scatter through $90°$ in the laboratory system is about the same as the time to scatter through $90°$ in the CM system, except when $m_1 \gg m_2$, in which case the lab $90°$ deflection time is about m_1/m_2 times longer than the CM $90°$ deflection time. Thus, the $90°$ deflection times in the lab system for ions (i) and electrons (e) are

$$\left(\tau_{90}^{ee}\right)_L = \frac{6\sqrt{6\pi}\sqrt{m_e}\varepsilon_0^2(kT)^{3/2}}{n_e(e^4)\ln\Lambda}$$

$$\left(\tau_{90}^{ii}\right)_L = \frac{6\sqrt{6\pi}\sqrt{m_i}\varepsilon_0^2(kT)^{3/2}}{n_i(e_i^4)\ln\Lambda}$$

$$\left(\tau_{90}^{ei}\right)_L = \frac{6\sqrt{3\pi}\sqrt{m_e}\varepsilon_0^2(kT)^{3/2}}{n_i(e_i^2e^2)\ln\Lambda}$$

$$\left(\tau_{90}^{ie}\right)_L = \frac{6\sqrt{3\pi}\sqrt{m_e}(m_i/m_e)\varepsilon_0^2(kT)^{3/2}}{n_e(e_i^2e^2)\ln\Lambda}$$

These times are in the ratio

$$\left(\tau_{90}^{ee}\right)_L : \left(\tau_{90}^{ii}\right)_L : \left(\tau_{90}^{ei}\right)_L : \left(\tau_{90}^{ie}\right)_L = 1 : \sqrt{\frac{m_i}{m_e}} : 1 : \frac{m_i}{m_e}$$

Thus, electrons lose their directed momentum equally fast to ions and to other electrons, while ions lose their directed momentum over a time longer by a factor $\sqrt{m_i/m_e}$ to other ions. In plasmas of fusion interest,

$$\left(\tau_{90}^{ee}\right)_L \simeq \left(\tau_{90}^{ei}\right)_L \simeq 10^{-4}\,\text{sec} \quad \left(\tau_{90}^{ii}\right)_L \simeq 10^{-2}\,\text{sec} \quad \left(\tau_{90}^{ie}\right)_L \simeq 1\,\text{sec}$$

The energy transferred from the incident particle 1 with energy E_0 to the target particle 2 is related to the scattering angle in the CM system by

$$\frac{\Delta E}{E_0} = \frac{4m_1m_2}{(m_1+m_2)^2}\sin^2\frac{\theta_c}{2}$$

a relationship which can be derived from conservation of energy and momentum. The energy transferred in a 90° deflection time in the CM system is found from this expression by setting $\theta_c = \pi/2$,

$$\left(\frac{\Delta E}{E_0}\right)_{90} \simeq \frac{2 m_1 m_2}{(m_1 + m_2)^2}$$

The number of 90° deflection times required to transfer essentially all the energy is

$$\frac{1}{(\Delta E/E_0)_{90}}$$

Thus, the characteristic energy transfer time is

$$\tau_E^{12} \simeq \frac{(\tau_{90}^{12})_c}{(\Delta E/E_0)_{90}}$$

Combining these results leads to

$$\tau_E^{ee} \simeq (\tau_{90}^{ei})_L$$

$$\tau_E^{ii} \simeq \sqrt{\frac{m_i}{m_e}} (\tau_{90}^{ei})_L$$

$$\tau_E^{ei} \simeq \frac{m_i}{m_e} (\tau_{90}^{ei})_L$$

$$\tau_E^{ie} \simeq \frac{m_i}{m_e} (\tau_{90}^{ei})_L$$

An energetic electron will share its energy with other electrons in a 90° deflection time for electrons. An energetic ion will share its energy with other ions over a time longer by a factor of $\sqrt{m_i/m_e}$. Electrons and ions exchange energy with each other over an even longer time, a factor of m_i/m_e greater than the electron 90° deflection time. From this it can be inferred that when the energy distributions of particles in a plasma are perturbed (e.g.. by external heating), the electrons will come to a new equilibrium by scattering among themselves – self-equilibrating – on a fast time scale ($\sim 10^{-4}$ sec in fusion plasmas), next the ions will self-equilibrate on a somewhat slower time scale ($\sim 10^{-2}$ sec), and finally the ions and electrons will exchange energy on the slowest time scale ($\sim 10^{-1}$ sec).

2.6
Resistivity

Now consider the acceleration of electrons under the action of an electric field, including the effect of collisions in destroying the ordered momentum created by the electric field. The equation of motion for an electron is

$$m_e \dot{v}_e = -e\mathbf{E} - \frac{m_e v_e}{(\tau_{90}^{ei})_L}$$

Writing $\mathbf{j} = -en_e\mathbf{v}_e$, this becomes

$$\frac{m_e}{e^2 n_e}\frac{d\mathbf{j}}{dt} = \mathbf{E} - \eta\mathbf{j}$$

where

$$\eta \equiv \frac{e_i e\sqrt{m_e}\ln\Lambda}{6\sqrt{3}\pi\varepsilon_0^2(kT)^{3/2}}$$

is the plasma resistivity. At steady state the equation of motion reduces to the familiar Ohm's law

$$\mathbf{E} = \eta\mathbf{j}$$

2.7
Gyromotion

The equation of motion for a charged particle moving in a uniform magnetic field is

$$m\dot{\mathbf{v}} = \mathbf{F} = e\mathbf{v} \times \mathbf{B}$$

Choosing the field in the z direction, the three components of this equation are

$$m\dot{v}_z = 0$$
$$m\dot{v}_x = eBv_y$$
$$m\dot{v}_y = -eBv_x$$

The implications of these equations are (1) that the magnetic field does not affect the motion along the magnetic field direction, but (2) that the magnetic field forces produce a circular motion about the field lines in the plane perpendicular to the magnetic field direction. Thus, the particles move along the magnetic field and spiral about it.

These equations have the formal solution

$$\mathbf{v} = v_{z0}\mathbf{n}_z + v_\perp\cos(\Omega t + \alpha)\mathbf{n}_x - v_\perp\sin(\Omega t + \alpha)\mathbf{n}_y$$

where

$$\Omega = \frac{eB}{m}$$

is the gyrofrequency and $v_\perp$ is the speed of circular motion (Figure 2.6).

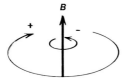

Figure 2.6 Gyromotion.

A charged particle will move along the magnetic field with whatever velocity it had initially, but its motion in the plane perpendicular to the magnetic field is circular. Thus, a charged particle spirals about a magnetic field line with gyrofrequency Ω and gyroradius

$$r_L \equiv \frac{v_\perp}{|\Omega|} = \frac{mv_\perp}{|e|B} \simeq \frac{mv_{th}}{|e|B}$$

where the thermal velocity is $v_{th} = \sqrt{2T/m}$.
  For ions (i) and electrons (e),

$$\Omega_e = -1.76 \times 10^{11} B(T) \, \text{rad/sec}$$
$$\Omega_i = 0.96 \times 10^{8} B(T) \, \text{rad/sec}$$

with B in Tesla and

$$r_{Le} = 1.066 \times 10^{-4} \frac{\sqrt{T_e(\text{keV})}}{B(T)} \, \text{m}$$

$$r_{Li} = 4.57 \times 10^{-3} \frac{\sqrt{A_i T_i(\text{keV})}}{z_i B(T)} \, \text{m}$$

where A_i and z_i are atomic mass (amu) and atomic number.

2.8
Drifts

When an electric field is present or when the magnetic field is nonuniform, the motion is more complicated because the form of the force term in the equation of motion is more complex. Fortunately the equation of motion is linear, so the motions arising from the various forces can be determined separately and superimposed. The motions due to the presence of non-uniformities in the magnetic field and to electric fields are referred to as drifts.

  Consider the case when an electric field is perpendicular to the uniform magnetic field. The particle equation of motion is

$$m\dot{\mathbf{v}} = e(\mathbf{E} + \mathbf{v} \times \mathbf{B})$$

  Writing

$$\mathbf{v} = \mathbf{v}_\Omega + \mathbf{v}_{\mathbf{E}\times\mathbf{B}}$$

where $\mathbf{v}_\Omega$ is the gyromotion, the equation of motion can be decomposed into two equations,

$$m\dot{\mathbf{v}}_\Omega = e\mathbf{v}_\Omega \times \mathbf{B} \quad (\text{gyromotion})$$
$$0 = e(\mathbf{E} + \mathbf{v}_{\mathbf{E}\times\mathbf{B}} \times \mathbf{B}) \quad (\text{E} \times \text{B drift})$$

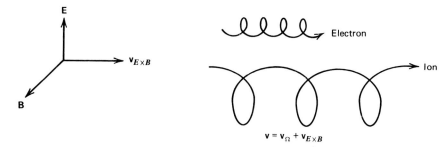

Figure 2.7 E × B drift.

From the second of these equations,

$$v_{E \times B} = \frac{E \times B}{B^2}$$

See Figure 2.7.

The $E \times B$ drift is in the same direction for ions and electrons. It produces a net motion of the plasma, but no net current or charge separation.

Now consider the situation when the magnetic field line is curved. It is well known from classical mechanics that a particle moving with speed $v_\parallel$ along a curved arc with radius of curvature R must experience a centrifugal force

$$F_c = \frac{-mv_\parallel^2}{R} \mathbf{n}_c$$

acting toward the center of curvature. ($\mathbf{n}_c$ is a unit vector in the direction of the radius of curvature).

Writing $\mathbf{v} = \mathbf{v}_\Omega + \mathbf{v}_c$, the equation of motion can be separated into an equation describing the gyromotion and an equation describing the curvature drift,

$$0 = \frac{-mv_\parallel^2}{R} \mathbf{n}_c + e\mathbf{v}_c \times \mathbf{B}$$

from which

$$\mathbf{v}_c = -\frac{mv_\parallel^2}{eR} \frac{\mathbf{B} \times \mathbf{n}_c}{B^2}$$

This curvature drift is orthogonal to the plane containing the magnetic field and field curvature vectors – out of the page for ions and into the page for electrons in Figure 2.8. This drift causes charge separation and a net current

$$\mathbf{j} = n_i e_i \mathbf{v}_{ci} + n_e e_e \mathbf{v}_{ce}$$

Finally, consider a nonuniform magnetic field strength, which produces a force

$$\mathbf{F} = -\frac{1}{2} \frac{mv_\perp^2}{B} \nabla B$$

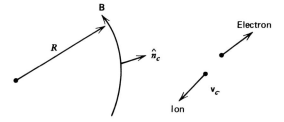

Figure 2.8 Curvature drift.

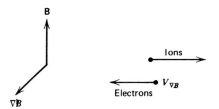

Figure 2.9 ∇**B** drift.

on a charged particle. Again separating the motion $\mathbf{v} = \mathbf{v}_\Omega + \mathbf{v}_{\nabla B}$, the equation of motion can be separated into an equation describing the gyromotion and an equation describing the ∇B drift,

$$0 = -\frac{\frac{1}{2} m v_\perp^2}{B} \nabla B + e\left(\mathbf{v}_{\nabla B} \times \mathbf{B}\right)$$

from which

$$\mathbf{v}_{\nabla B} = \frac{m v_\perp^2}{2e} \frac{\mathbf{B} \times \nabla \mathbf{B}}{B^3}$$

See Figure 2.9.

The ions and electrons drift in opposite directions, producing a net current and a charge separation.

In all cases considered, the drift velocity is perpendicular to the force causing the drift and to the magnetic field.

Problems

2.1. Calculate the Debye length for a 15-keV plasma with an electron density of $2 \times 10^{20}\,\mathrm{m}^{-3}$. Repeat for a 100-eV plasma with an electron density of $1 \times 10^{19}\,\mathrm{m}^{-3}$.

2.2. Calculate the electrostatic potential from a point charge located at $r = 0$.

2.3. Calculate the electron and proton plasma frequencies in a hydrogen plasma with an electron density of $10^{20}\,\mathrm{m}^{-3}$.

2.4. Derive the relationship given in the text between the scattering angles in the lab and the CM system from conservation of kinetic energy and conservation of momentum.

2.5. For the same value of the impact parameter, would the deflection of electrons be greater for scattering from hydrogen or helium?

2.6. Calculate the $90°$ lab deflection times for protons scattering on protons and on electrons and for electrons scattering on protons and on electrons, for a plasma with a temperature of $10\,\text{keV}$ and an electron density of $2 \times 10^{20}\,\text{m}^{-3}$.

2.7. Repeat Problem 2.6 for a plasma with a temperature of $50\,\text{eV}$ and a density of $2 \times 10^{19}\,\text{m}^{-3}$.

2.8. Derive from conservation of energy and momentum the relation given in the text between the energy transferred in a Coulomb scattering interaction and the scattering angle in the CM system.

2.9. A very energetic electron is injected into a 10-keV hydrogen plasma. Will it lose, via collisions, most of its energy to plasma ions or plasma electrons?

2.10. An external electric field of $E = 0.1\,\text{V/m}$ is imposed on a hydrogen plasma with density $5 \times 10^{19}\,\text{m}^{-3}$ and temperature $100\,\text{eV}$. Calculate the resulting steady-state current density in the plasma. Repeat for a temperature of $10\,\text{keV}$.

2.11. Calculate the electron and proton gyroradii and gyrofrequencies for the plasma parameters of Problem 2.1 and a magnetic field of 5 T. Repeat for a magnetic field of 0.5 T.

2.12. Calculate the $E \times B$ drift speed in a plasma at $10\,\text{keV}$ with $E = 0.1\,\text{V/m}$ and $B = 5\text{T}$. Compare this with the thermal velocity.

2.13. Calculate the curvature drift speed for protons and electrons in plasma at $10\,\text{keV}$ with $B = 5\,\text{T}$ and a field radius of curvature $R = 10\,\text{m}$.

2.14. Derive an expression for the drift velocity due to gravity.

3
Equilibrium and Transport

3.1
Equilibrium and Pressure Balance

The basic equations which determine the plasma equilibrium are the momentum balance equation relating the force due to the pressure gradient to the Lorentz force acting on all of the particles in a unit volume of plasma.

$$0 = -\nabla p + \mathbf{j} \times \mathbf{B}$$

and Ampere's law

$$\mu_0 \mathbf{j} = \nabla \times \mathbf{B}$$

From the first equation, $\mathbf{j} \cdot \nabla p = 0$ and $\mathbf{B} \cdot \nabla p = 0$. Thus, the current and the magnetic field lie in isobaric surfaces, known as flux surfaces. Note that the current and the magnetic field are not parallel, except under the special condition $\nabla p = 0$.

As an example, consider a cylindrical plasma column carrying a uniform axial current [$j_z(r) = j_{z0} =$ constant], as depicted in Figure 3.1. If there is axial and poloidal (θ) symmetry, the above two equations reduce to

$$\frac{dp}{dr} = -j_z B_\theta$$

and

$$\mu_0 j_z = \frac{1}{r} \frac{d}{dr} (r B_\theta)$$

which can be solved, for $j_z(r) = j_{z0}$, to obtain

$$B(r) = \begin{cases} \dfrac{\mu_0}{2} j_{z0} r, & r \le a \\[2mm] \dfrac{\mu_0}{2} j_{z0} \dfrac{a^2}{r}, & r > a \end{cases}$$

Fusion: Second, Completely Revised and Enlarged edition. Weston M. Stacey
Copyright © 2010 WILEY-VCH Verlag GmbH & Co. KGaA, Weinheim
ISBN: 978-3-527-40967-9

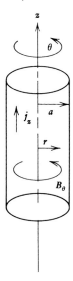

Figure 3.1 Cylindrical plasma column.

and

$$p(r) = \frac{1}{4}\mu_0 j_{z0}^2 (a^2 - r^2) \quad r < a$$

The isobaric surfaces are $r =$ constant annuli. The current, j_z and the magnetic field B_θ lie in the annuli, that is, there is no r-component of either **j** or **B**.

When a long cylinder is bent (figuratively) to form a torus, the flux surfaces remain, to a good approximation, nested annuli, but now with a small deformation and displacement due to the toroidicity. The flux surfaces become somewhat more complicated for other magnetic geometries, but retain the general property of being nested surfaces that are more or less "parallel" to the outer surface of the plasma.

Since charged particles follow magnetic field lines and the latter lie in isobaric surfaces, charged particles move freely on flux surfaces but are constrained from moving across flux surfaces. Actually, it is the center about which the particle gyrotates that moves on the flux surface; the particle itself makes small excursions on the order of the gyroradius from the flux surface.

Combining the first two equations yields

$$\nabla\left(p + \frac{B^2}{2\mu_0}\right) = \frac{1}{\mu_0}(\mathbf{B}\cdot\nabla)\mathbf{B}$$

The RHS vanishes when the field lines are straight and is small in most cases of interest, leading to the useful approximate result

$$p + \frac{B^2}{2\mu_0} \simeq \text{constant}$$

namely, that the sum of the kinetic and magnetic pressures is constant.

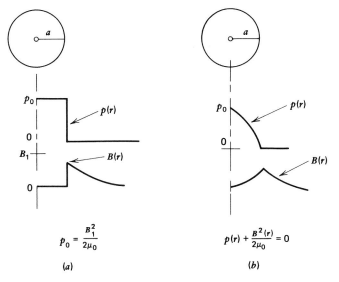

Figure 3.2 Pressure balance in plasma cylinder.

As an idealized example, consider a plasma column in which the kinetic pressure is uniform for $r < a$ and the axial current all flows just on the surface at $r = a$, in which case the magnetic field changes discontinuously from zero at $r = a - \varepsilon$ to B_1 at $r = a + \varepsilon$, as depicted in Figure 3.2a. A more realistic situation, in which the pressure decreases gradually with radius and the current is distributed over the plasma cross section, is depicted in Figure 3.2b.

3.2
Classical Transport

In the previous section it was shown that particles move freely on nested flux surfaces that are "parallel" to the outer surface of the plasma. Now it will be established that particles also move across flux surfaces because of collisions.

Consider again the cylindrial plasma depicted in Figure 3.1, but now with an axial field and the possibility of poloidal currents. The flux surfaces are still nested annuli.

A simple picture of transport across flux surfaces results from considering that since a particle has excursions from a flux surface on the order of its gyroradius, a series of small-angle collisions can displace the center of the gyro orbit from one flux surface to another. This displacement can be inward or outward, so the net displacement must be treated as a random walk process governed by a diffusion coefficient,

$$D = \frac{\overline{(\Delta x)^2}}{\tau}$$

where $\overline{(\Delta x)^2} = r_L^2$ is the mean-square characteristic displacement and $\tau = (\tau_{90})_L$ is the characteristic time for that displacement. Thus, using a Fick's law-type diffusion model, the outward radial transport across flux surfaces should be

$$n v_r \simeq - \left[\frac{r_L^2}{(\tau_{90})_L} \right] \frac{\partial n}{\partial r}$$

This result will now be derived more formally.

Consider the momentum balance equation on all the particles of type i (e.g., ions or electrons),

$$0 = -\nabla p_i + n_i e_i (-\nabla \Phi + \mathbf{v}_i \times \mathbf{B}) + \mathbf{R}_i$$

where $\mathbf{R}_i$ is the momentum exchange due to collisions between particles of species i and particles of all other species, which can be approximated

$$\mathbf{R}_i = - \sum_{j \neq i} \frac{n_i m_i}{(\tau_{90}^{ij})_L} (\mathbf{v}_i - \mathbf{v}_j)$$

(Note that summing the above momentum balance over all species yields the momentum balance on the plasma that was used in the previous section because $\sum_i \mathbf{R}_i = 0$ from conservation of momentum and $\sum_i n_i e_i = 0$ from charge neutrality. Note further that collisions among particles of the same species do not result in any net momentum gain or loss by that species, hence do not enter into the momentum balance.)

The radial component of the momentum balance equation is

$$\left(\frac{\partial p_i}{\partial r} + n_i e_i \frac{\partial \Phi}{\partial r} \right) = n_i e_i (v_{i\theta} B_z - v_{iz} B_\theta) + R_{ir}$$

The collisional term R_{ir} involves particle flows across flux surfaces, and thus is small compared to the terms involving flows in the flux surfaces. Imposing the condition $B_z \gg B_\theta$ this equation yields

$$n_i v_{i\theta} = \frac{1}{e_i B_z} \left(\frac{\partial p_i}{\partial r} + n_i e_i \frac{\partial \Phi}{\partial r} \right)$$

The poloidal component of the momentum balance equation is

$$n_i e_i v_{ir} B_z + R_{i\theta} = \frac{1}{r} \frac{\partial p_i}{\partial \theta} + n_i e_i \frac{1}{r} \frac{\partial \Phi}{\partial \theta} = 0$$

where the RHS vanishes from symmetry considerations.

Combining these results, yields an expression for the radial particle flow,

$$n_i v_{ir} = \frac{R_{i\theta}}{e_i B_z} = - \frac{n_i m_i}{e_i B_z^2} \sum_{j \neq i} (\tau_{90}^{ij})_L^{-1} \left(\frac{1}{n_i e_i} \frac{\partial p_i}{\partial r} - \frac{1}{n_j e_j} \frac{\partial p_j}{\partial r} \right)$$

Thus, the radial particle flow is driven by the poloidal component of the collisional momentum exchange, which in turn is dependent upon the differences in radial pressure gradients among the species. For the case of $T_i = $ constant the previous equation has the form

$$n v_r \simeq -\frac{mkT}{e^2 B_z^2 (\tau_{90})_L} \frac{\partial n}{\partial r}$$

$$\simeq -\frac{m(mv_{th}^2)}{e^2 B_z^2 (\tau_{90})_L} \frac{\partial n}{\partial r} \simeq -\frac{r_L^2}{(\tau_{90})_L} \frac{\partial n}{\partial r}$$

which is the result obtained previously from the simple random walk argument and Fick's law.

Specializing the above result to a plasma consisting of only one ($z = 1$) ion species i and electrons e and writing $p = p_i + p_e$ leads to

$$n_i v_{ir} = -\frac{m_i (\tau_{90}^{ie})_L^{-1}}{e^2 B_z^2} \frac{\partial p}{\partial r}$$

and

$$n_e v_{er} = -\frac{m_e (\tau_{90}^{ei})_L^{-1}}{e^2 B_z^2} \frac{\partial p}{\partial r}$$

Since $n_e m_e (\tau_{90}^{ei})_L^{-1} = n_i m_i (\tau_{90}^{ie})_L^{-1}$ (see Section 2.5)

$$n_e v_{er} = n_i v_{ir}$$

and the radial current vanishes,

$$j_r \equiv e n_i v_{ir} + (-e) n_e v_{er} = 0$$

Vanishing of the radial current is known as ambipolarity and is necessary to ensure charge neutrality.

Now, consider a plasma that also contains a $z > 1$ ion species z. The collisional momentum exchange term for the $z = 1$ ions i is

$$\mathbf{R}_i = -n_i m_i \left[\left(\tau_{90}^{ie} \right)_L^{-1} (\mathbf{v}_i - \mathbf{v}_e) + \left(\tau_{90}^{iz} \right)_L^{-1} (\mathbf{v}_i - \mathbf{v}_z) \right]$$

Recalling from Section 2.5 that

$$\left(\tau_{90}^{iz} \right)_L^{-1} = \frac{n_z z^2}{n_e} \sqrt{\frac{m_i}{m_e}} \left(\tau_{90}^{ie} \right)_L^{-1}$$

it is apparent that collisions with electrons are unimportant relative to collisions with impurity ions when

$$\frac{n_z z^2}{n_e} > \sqrt{\frac{m_e}{m_i}} \simeq \frac{1}{50}$$

in which case the radial particle fluxes may be written as

$$n_i v_{ir} = -\frac{n_i m_i (\tau_{90}^{iz})_L^{-1}}{e^2 B_z^2} \left(\frac{1}{n_i} \frac{\partial p_i}{\partial r} - \frac{1}{zn_z} \frac{\partial p_z}{\partial r} \right)$$

$$n_z v_{zr} = -\frac{n_z m_z (\tau_{90}^{zi})_L^{-1}}{ze^2 B_z^2} \left(\frac{1}{zn_z} \frac{\partial p_z}{\partial r} - \frac{1}{n_i} \frac{\partial p_i}{\partial r} \right)$$

$$n_e v_{er} = n_i v_{ir} + zn_z v_{zr}$$

where the last result is required to maintain charge neutrality.

Noting that $n_i m_i (\tau_{90}^{iz})_L^{-1} = n_z m_z (\tau_{90}^{zi})_L^{-1}$, it follows that the condition for a steady-state solution (with no internal particle sources), $n_i v_{ir} = 0$, requires

$$\frac{1}{n_i} \frac{\partial p_i}{\partial r} = \frac{1}{zn_z} \frac{\partial p_z}{\partial r}$$

Ignoring temperature gradients, this condition implies the solution

$$\frac{n_z(r)}{n_z(0)} = \left[\frac{n_i(r)}{n_i(0)} \right]^z$$

Thus, the impurity distribution would be much more sharply peaked toward the center than the ion distribution if $z \gg 1$. This situation would lead to large radiative power losses, which would adversely affect the power balance in the plasma and possibly cool the plasma prematurely. (Temperature gradient effects, which have not been considered here, mitigate this central impurity peaking problem.)

An analogous procedure can be used to develop expressions for heat conduction, starting with the heat flux balance equation. The characteristic parameter which governs the heat conduction across flux surfaces is again proportional to $r_L^2 (\tau_{90})_L^{-1}$.

Curvature and non-uniformity of the magnetic field modify the simple transport relations of this section when confinement geometries other than the simple cylinder are considered. "Neoclassical" transport theory results from taking these effects into account.

3.3
Neoclassical Transport

When the derivation of the previous section is carried out in toroidal (rather than cyclindrical) geometry, it is no longer possible to make use of poloidal symmetry to justify setting poloidal gradients of pressure and potential to zero. Instead, these terms must be evaluated from the parallel component of the momentum balance equation. When this is done, and the friction term is generalized to include fitted numerical coefficients along with a thermal friction term, arising from the temperature dependence of the collision frequency, the resulting expression for the radial particle flux averaged over the poloidal angle is

$$\Gamma_{\sigma r}^{PS} \equiv \langle n_\sigma v_{\sigma r}^{PS} \rangle \simeq - \frac{2\varepsilon^2 n_\sigma m_\sigma v_{\sigma\sigma'}}{\left(e_\sigma B_\theta^0\right)^2}$$

$$\times \left[\left(C_1 + \frac{C_2^2}{C_3} \right) \left(\frac{1}{n_\sigma} \frac{\partial p_\sigma}{\partial r} - \frac{1}{z n_{\sigma'}} \frac{\partial p_{\sigma'}}{\partial r} \right) - \frac{5}{2} \frac{C_2}{C_3} \frac{\partial T_\sigma}{\partial r} \right],$$

where $z \equiv \frac{e_{\sigma'}}{e_\sigma}$ and $\sigma = i, z, e$.

Comparing this expression with the above expression for the classical particle flux in a cyclinder, and noting that the constants C_1, C_2, C_3 are of order unity, leads to an estimate of the relative magnitudes of these Pfirsch-Schluter and the classical radial particle fluxes

$$\Gamma_{\sigma r}^{PS} \simeq 2 \left(\frac{\varepsilon B_\phi^0}{B_\theta^0} \right)^2 \Gamma_{\sigma r}^C = 2q^2 \Gamma_{\sigma r}^C.$$

Since q varies from ~ 1 at the plasma center to 3–4 at the edge, we see that the Pfirsch-Schluter fluxes are an order of magnitude larger than the classical fluxes. Thus, these toroidal geometry effects, which are associated with the grad-B and curvature drifts introduced by the toroidal geometry, result in an order-of-magnitude enhancement of the radial transport fluxes over the classical values.

The fact that the field in a tokamak is helical, rather that straight as in the cylindrical model, introduces a further effect on transport. The helical field introduces a magnetic well in the B-field on the outboard side of the torus and a maximum in the B-field on the inboard side of the torus; that is, a B-field configuration such as is shown in Figure 4.1 for the mirror plasma. Following a similar development as for the mirror confinement, it can be shown that a fraction $\simeq \sqrt{r/R} \equiv \sqrt{\varepsilon}$ of the particles are trapped in this magnetic well, producing a radial particle flux

$$\Gamma_{\sigma r}^{\text{trap}} \simeq \frac{1}{2} \varepsilon^{-\frac{3}{2}} \Gamma_{\sigma r}^{PS} = \varepsilon^{-\frac{3}{2}} q^2 \Gamma_{\sigma r}^C$$

that is an order of magnitude larger than the Pfirsch-Schluter flux due to drifts and is two orders of magnitude larger than the classical particle flux in cylindrical geometry. This trapped particle flux only exists when the particles have time to execute a trapped particle orbit before being scattered through $90°$, which requires that $qR/v_{th}\tau_{90} < \varepsilon^{\frac{3}{2}}$.

3.4
Fluctuation-Driven Transport

The neoclassical transport described in the previous section sets a lower limit on particle (and energy) transport in a tokamak. In practice, the transport rate inferred from measured energy confinement times is almost always larger than the rate that would be inferred from neoclassical theory. This enhanced transport is believed to be due to the turbulent fluctuations in the density, electrostatic potential and magnetic field that are commonly measured in tokamaks and other plasmas.

There are two obvious ways that turbulence can lead to enhanced radial particle and energy transport: (1) the fluctuating electrostatic potential can produce a fluctuating electric field which drives fluctuating ExB particle drifts; and (2) a fluctuating radial component of the magnetic field can cause a fluctuating radial component of the large particle flux parallel to the field lines. Various theoretical descriptions of transport rates corresponding to different turbulent instabilities have been derived, but to date this subject is far from understood.

Problems

3.1. Calculate the radial pressure distribution and magnetic field distribution in a long plasma cylinder that carries an axial current distribution $j_z(r) = 10^2$ $(1 - r/a)$A/cm^2.

3.2. Calculate the magnetic pressure associated with a field $B = 6$ T.

3.3. Calculate the radial ion transport flux $n_i v_{ir}$ in a cylindrical hydrogen plasma which has an axial current of 1 MA uniformly distributed over a plasma of radius $a = 1$ m, has a uniform temperature of $T_i = T_e = 2$ keV, and has an axial magnetic field $B_z = 5$ T. Use $n = 10^{20}$ m^{-3} and ln $\Lambda = 20$.

3.4. Discuss all the drifts of ions and electrons for the cylindrical plasma with a uniform current density that was introduced in Section 2.8. What is the effect of these drifts upon confinement?

4
Confinement Concepts

The two principal types of confinement concepts being investigated throughout the world – open (mirror) and closed (toroidal) – are described in this chapter. The leading and alternative concepts are discussed briefly.

4.1
Magnetic Mirrors

The principles of confinement in magnetic wells has evolved through several stages, as summarized in Section 1.2.2. In this section the principles of mirror confinement will be developed.

4.1.1
Simple Mirrors

The simple mirror configuration is illustrated in Figure 4.1. Conservation of kinetic energy requires

$$\frac{1}{2}m\left[v_\parallel^2(s) + v_\perp^2(s)\right] = \mathrm{KE} = \text{constant}$$

since the stationary magnetic field does no work on a charged particle. Conservation of angular momentum requires

$$p_\theta = mr_L v_\perp = \frac{mv_\perp^2(s)}{B(s)} = \text{constant}$$

which is conventionally written

$$\mu = \frac{\frac{1}{2}mv_\perp^2(s)}{B(s)} = \text{constant}$$

Combining these relationships yields

$$\frac{1}{2}mv_\parallel^2(s) = \mathrm{KE} - \mu B(s)$$

Fusion: Second, Completely Revised and Enlarged edition. Weston M. Stacey
Copyright © 2010 WILEY-VCH Verlag GmbH & Co. KGaA, Weinheim
ISBN: 978-3-527-40967-9

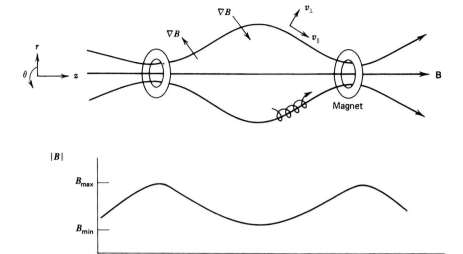

Figure 4.1 Simple mirror configuration.

indicating that particles for which $KE/\mu = B(s)$ for $B_{\min} \le B(s) \le B_{\max}$ will be trapped; that is, $v_\parallel$ will vanish.

Evaluating the constants KE and μ from the defining relationships at $s = s_0$ yields

$$\frac{1}{2}mv_\parallel^2(s) = \left(\frac{1}{2}mv_\parallel^2(s_0) + \frac{1}{2}mv_\perp^2(s_0)\right) - \frac{1}{2}mv_\perp^2(s_0)\frac{B(s)}{B_{\min}}$$

Thus, the condition for a particle to be trapped depends upon $v_\perp(s_0)/v_\parallel(s_0)$ and $B(s)/B_{\min}$.

The boundary in velocity $(v_\parallel, v_\perp)$ space between trapped and untrapped particles can be determined by evaluating the above equation for $v_\parallel(s_{\max}) = 0$,

$$v_\perp(s_0) = \pm\left(\frac{B_{\max}}{B_{\min}} - 1\right)^{-1/2} v_\parallel(s_0)$$

Noting that $v_\perp$ is the two-dimensional velocity component in the plane perpendicular to the magnetic field, this equation defines a cone, as depicted in Figure 4.2.

The magnetic field in a simple mirror has both curvature and gradient. Thus, curvature and ∇B drifts will be present. Using the expressions of Section 2.8 and the (r, θ, z) coordinate system shown in Figure 4.1, these drifts are seen to be in the θ direction,

$$\mathbf{v}_c = -\frac{mv_\parallel^2}{eR_c}\frac{\mathbf{B} \times \mathbf{n}_c}{B^2} = \frac{mv_\parallel^2}{eR_c B}\mathbf{n}_\theta$$

$$\mathbf{v}_{\nabla B} = \frac{\frac{1}{2}mv_\perp^2}{e}\frac{\mathbf{B} \times \nabla B}{B^3} = \frac{\frac{1}{2}mv_\perp^2}{e}\frac{|\nabla B|}{B^2}\mathbf{n}_\theta$$

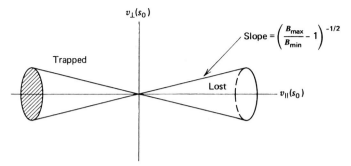

Figure 4.2 Loss cone.

resulting in rotation about the axis of symmetry, but no net radial motion. Hence, the drifts do not destroy confinement.

Particles with $v_\perp(s_0)/v_\parallel(s_0)$ which fall within the loss cone shown in Figure 4.2 are lost immediately. Other particles scatter into the loss cone and then are lost. Thus, the confinement time is proportional to the time required for a particle to scatter into the loss cone. This time is related to the 90° deflection time in the lab, $(\tau_{90})_L$,

$$\tau_p = (\tau_{90})_L \log_{10}\left(\frac{B_{max}}{B_{min}}\right)$$

Since $(\tau_{90}^{ii})_L \simeq \sqrt{m_i/m_e}(\tau_{90}^{ee})_L$, the electrons scatter into the loss cone and escape much faster than do the ions. This creates a positive net charge in the plasma which results in a positive electrostatic potential that acts to confine electrons in the loss cone until ions escape. Thus, the particle confinement time for ions and electrons in a simple mirror is proportional to the ion 90° deflection time,

$$\tau_p \simeq (\tau_{90}^{ii})_L \log_{10}\left(\frac{B_{max}}{B_{min}}\right)$$

The plasma density is maintained by reinjecting ions at somewhat higher energy.

The simple mirror is unstable against flute-type radial perturbations of the plasma. Consider the idealized force balance on the plasma surface illustrated in Figure 4.3. The internal kinetic pressure p exerts an outward force which is balanced in equilibrium by the inward force of the magnetic pressure, $B^2/2\mu_0$. Now imagine that the surface is perturbed as indicated by the dashed line in Figure 4.3 outward into a region of weaker magnetic field. The outward force of the kinetic pressure remains the same, but now the inward force of the magnetic pressure is reduced, leading to further surface deformation and ultimately to destruction of confinement.

4.1.2
Minimum-*B* Mirrors

The flute instability can be suppressed if the field increases, rather than decreases, away from the plasma, that is, if the plasma sits in a three-dimensional magnetic well.

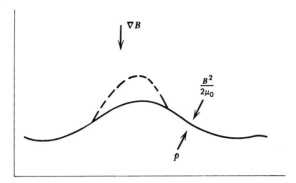

Figure 4.3 Flute instability.

In this case the deformation of Figure 4.3 results in the inward force of the magnetic pressure being stronger than the outward force of the kinetic pressure, thus acting to restore the surface to its equilibrium position.

Minimum-B stabilization of mirror plasmas has evolved through a number of magnetic configurations. The final stage of this evolution makes use of a set of "yin-yang" coils wound somewhat like the seams of a baseball (Figure 4.4).

The principles of confinement in a minimum-B mirror are essentially the same as those described previously for the simple mirror, namely, confinement is governed by ions scattering into the loss cone. The estimate given previously for confinement time is still valid. The drifts are considerably more complicated, but it is possible to show that they do not result in particle motion out of the confinement region.

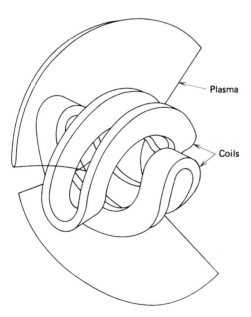

Figure 4.4 Schematic of mirror confinement configuration with yin–yang magnetic coils.

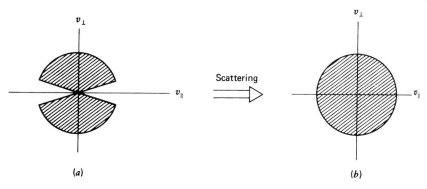

Figure 4.5 Relaxation of loss cone velocity distribution.

The basic minimum-B mirror is unstable against "loss-cone" instabilities driven by the relaxation of the anisotropic velocity distribution. The process is illustrated in Figure 4.5. Scattering events tend to cause the distribution of Figure 4.5a to relax toward that of Figure 4.5b, which has less free energy. The difference in free energy is converted into plasma kinetic energy during the relaxation process, thus constituting an instability which destroys confinement. This loss-cone instability can be suppressed by filling in the distribution of Figure 4.5a by injecting ions with $v_\parallel / v_\perp \gg 1$.

The minimum-B mirror has been thoroughly studied and is reasonably well understood. The most favorable performance projected for a minimum-B mirror leads to a plasma power amplification factor of $Q_p \simeq 1-2$. Recalling that $Q_p \gtrsim 15$ is required for positive power balance on a reactor, it follows that the reactor prospects for a minimum-B mirror reactor for electricity production are poor. The minimum-B mirror has potential applications as a neutron source for materials testing or producing fissile material by nuclear transmutation of thorium-232 or uranium-238.

4.1.3
Tandem Mirrors

The most promising use of minimum-B mirrors, however, is as end plugs to confine ions electrostatically in a central solenoidal cell. (Recall that the simple and minimum-B mirrors operate with a slight positive charge because electrons deflect into the loss cone faster than ions).

The basic idea of the tandem mirror, illustrated in Figure 4.6, is to create a potential difference ϕ_i between the end plugs and the central cell by creating a density difference. According to the Maxwell–Boltzmann distribution

$$n_e(z) = n_0 \exp\left(\frac{e\phi}{kT_e}\right)$$

Using the subscripts p and c to refer to the end plug and the central cell, respectively, this relation can be used to obtain

Low–field solenoid

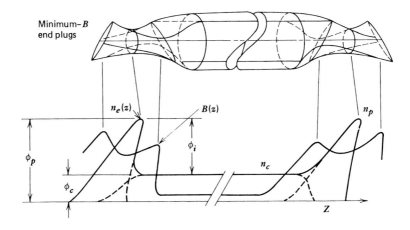

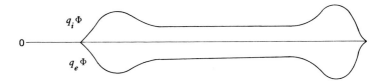

Figure 4.6 Tandem mirror with minimum-B end plugs. (Reproduced with permission of University of California, Lawrence Livermore National Laboratory, and the Department of Energy.)

$$e\phi_i \equiv e(\phi_p - \phi_c) = kT_e \ln\left(\frac{n_p}{n_c}\right) = kT_e \ln\left(\frac{\beta_p B_p^2}{\beta_c B_c^2}\right)$$

where the definition

$$\beta \equiv \frac{nk\,T}{B^2/2\mu_0}$$

has been used.

The basic tandem mirror concept has been established experimentally. Since most of the plasma volume in the tandem mirror is in the central cell, the power density $[\propto n^2\langle\sigma v\rangle]$ is proportional to $(\beta_c B_c^2)^2$. Thus, in order to achieve large confining potential and an adequate power density, it is necessary to have a large $\beta_p B_p^2$. Calculations indicate that magnetic fields in excess of 20T may be required in the end plugs, which would put a severe demand upon superconducting magnet technology. In addition, penetration of the dense end plug plasmas would require neutral beam energies on the order of 1 MeV, again placing a severe demand upon the technology.

A reduction of technological requirements can be achieved, in principle, by the addition of two thermal barrier cells. The original idea of the thermal barrier is

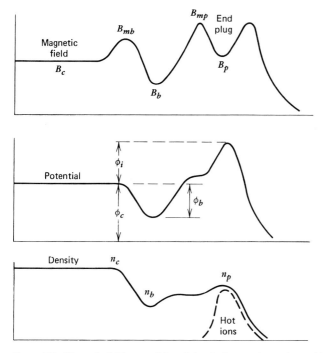

Figure 4.7 Magnetic field, potential, and density in a tandem mirror with thermal barrier. (Reproduced with permission of University of California, Lawrence Livermore National Laboratory, and the Department of Energy.)

shown in Figure 4.7. It is produced by adding a simple mirror (the barrier coil) at each end of the solenoid, thus partially isolating the end plugs from the solenoid. The intention is to insulate hotter electrons thermally in the end plug from relatively colder electrons in the central cell. If a higher electron temperature can be sustained in the plugs by intense electron heating (e.g., with microwaves at the electron-cyclotron frequency), then the potential peak in the plug necessary to confine the ions escaping from the solenoid can be generated with a much lower density n_p in the end plugs. A large reduction in the plug density reduces the power consumed in the plugs (hence improving Q) and opens up options for less demanding magnet and neutral beam technology.

Electrons coming from the central solenoid into the barrier see a negative potential ϕ_b, which can be derived from the Maxwell–Boltzmann distribution

$$e\phi_b = kT_c \ln\left(\frac{n_c}{n_b}\right)$$

Electrons coming from the plug into the barrier see a negative potential $\phi_i + \phi_b$, given by

$$e(\phi_i + \phi_b) = kT_p \ln\left(\frac{n_p}{n_b}\right)$$

Combining these two relations yields

$$e\phi_i = kT_p \ln\left(\frac{n_p}{n_c}\right) + k(T_p - T_c) \ln\left(\frac{n_c}{n_b}\right)$$

Confinement in the central solenoid is described by

$$(n\tau_p)_c(\text{cm}^{-3}\text{sec}) \simeq 10^{11}(kT_c)^{1/2}\left(\frac{e\phi_i}{kT_c}\right)\exp\left(\frac{e\phi_i}{kT_c}\right)$$

where kT is expressed in kiloelectron volts. This is substantially larger than for single-cell mirrors when $e\phi_i/kT_c \simeq 2\text{-}3$.

4.1.4
Current Mirror Research

Through the 1980s there were major research programs in mirror confinement in the USA and Russia, but projections of the mirror concept to a power reactor were not favorable. Partly as a result of these projections, the mirror confinement program in the USA was phased out in the 1980s, leaving the Russian experiments and new Japanese and Korean experiments to carry forward the development of the concept. These efforts have achieved keV level plasma temperatures and energy confinement times in the 0.05s range with tandem mirror configurations. Researchers believe that the mirror concept has potential for a materials test facility.

4.2
Tokamaks

4.2.1
Confinement Principles

Tokamaks are a class of closed toroidal confinement concepts in which a strong toroidal field B_ϕ is produced in the plasma by a set of external toroidal field coils. The basic geometry is indicated in Figure 4.8, and two coordinate systems (r, θ, ϕ) and (R, z, ϕ) are defined in Figure 4.9. The toroidal field is produced by a current flowing in the toroidal field coils, and within the plasma region can be represented as

$$B_\phi(r, \theta) = \frac{B_\phi^0}{(1 + r/R_0)\cos\theta} \equiv \frac{B_\phi^0}{1 + \varepsilon\cos\theta}$$

where $B_\phi^0 = B_\phi(r = 0)$ and R_0 is the radius of the minor axis relative to the major axis.

Confinement is not possible with a purely toroidal field because the ∇B and curvature drifts

$$\mathbf{v}_{\nabla B} = \frac{\frac{1}{2}mv_\perp^2}{eB^3}\mathbf{B}\times\nabla B = \frac{\frac{1}{2}mv_\perp^2}{eB_\phi^2}\frac{\partial B_\phi}{\partial R}\mathbf{n}_z$$

$$\mathbf{v}_c = -\frac{mv_\parallel^2}{eB^2}\frac{\mathbf{B}\times\mathbf{n}_c}{R_c} = \frac{mv_\parallel^2}{eB_\phi R}\mathbf{n}_z$$

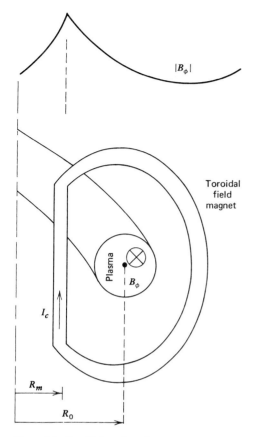

Figure 4.8 Toroidal configuration.

are up for positive ions and down for electrons for the direction of B_ϕ shown in Figure 4.8 (into the page). These drifts lead to charge separation, which produces a downward electric field, which in turn leads to a radially (R) outward $E \times B$ drift,

$$\mathbf{v}_{E \times B} = \frac{\mathbf{E} \times \mathbf{B}}{B^2} = \frac{E}{B_\phi} \mathbf{n}_c$$

that carries the plasma to the chamber wall. This is illustrated in Figure 4.10.

When a poloidal field is superimposed on the toroidal field, the radial (r) displacement due to the curvature and ∇B drifts is averaged out. $\mathbf{v}_{\nabla B}$ and $\mathbf{v}_c$ are still upward for ions, but now as the ion follows a field line which rotates poloidally as well as going around toroidally, an upward drift is radially outward for $0 < \theta < \pi$ and radially inward for $\pi < \theta < 2\pi$. The concept is illustrated in Figure 4.11.

The poloidal field can be produced by a set of external coils (the stellarator) or by toroidal current flowing in the plasma (the tokamak). In the tokamak, the poloidal field is related to the toroidal current density by Ampere's law,

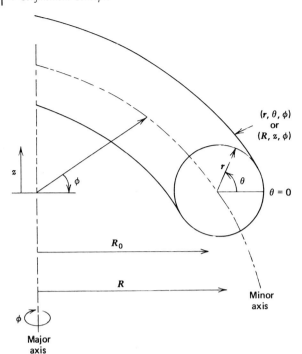

Figure 4.9 Toroidal (r, θ, ϕ) and cylindrical (R, z, ϕ) coordinate systems.

$$2\pi r B_\theta(r) = \int_0^r 2\pi r' j_\phi(r')dr'$$

The solenoidal condition $\nabla \cdot \mathbf{B} = 0$ can be written in toroidal coordinates as

$$\frac{R_0}{r} \frac{\partial}{\partial \theta}[1 + \varepsilon \cos\theta)B_\theta(r, \theta)] + \frac{\partial B_\phi(r, \theta)}{\partial \phi} = 0$$

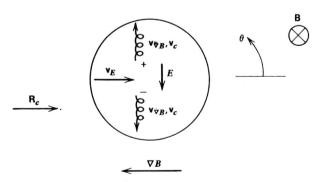

Figure 4.10 Curvature, and grad B, and $\mathbf{E} \times \mathbf{B}$ drifts in a simple toroidal system.

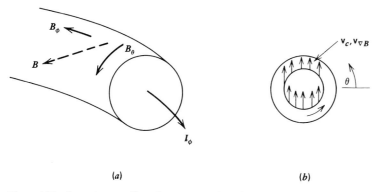

Figure 4.11 Averaging-out effect of curvature and ∇B drifts: (a) net field; (b) projection of poloidal motion.

The last term vanishes because of axisymmetry, leading to the result

$$B_0(r,\theta) = \frac{\text{constant}}{1 + \varepsilon \cos\theta} \equiv \frac{B_0^0}{1 + \varepsilon \cos\theta}$$

where $B_0^0 = B_0\left(r, \theta = \frac{\pi}{2}\right)$.

Thus, the total field can be written

$$B(r,\theta) = \frac{B_0}{1 + \varepsilon \cos\theta}$$

Now consider the pressure balance equilibrium on an idealized tokamak plasma with a uniform kinetic pressure and a surface current. The inward magnetic pressure balances, on the average, the outward kinetic pressure,

$$p = \frac{\langle B^2 \rangle}{2\mu_0}$$

Since the external surface is an isobaric surface, the kinetic pressure is uniform over the surface. However, the magnetic pressure varies over the surface, as indicated in Figure 4.12 and the above equation.

The magnetic force is larger than the kinetic pressure on the inboard at $\theta = \pi$, where

$$B(\theta = \pi) = \frac{B_0}{1 - \varepsilon}$$

and is smaller than the kinetic pressure on the outboard at $\theta = 0$, where

$$B(\theta = 0) = \frac{B_0}{1 + \varepsilon}$$

Another way to understand the force imbalance is to consider the Lorentz force of interaction between the plasma current I_ϕ and the poloidal field B_θ. The vertical component of the poloidal field is upward for $3\pi/2 < \theta < \pi/2$ and downward for $\pi/2 < \theta < 3\pi/2$ for a plasma current as shown in Figure 4.11. The downward

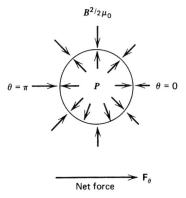

Figure 4.12 Force imbalance.

component interacts with the plasma current to produce an outward force, and the upward component interacts to produce an inward force. Since the downward component of the poloidal field is stronger, there is a net outward force $\mathbf{F}_\theta$.

This outward force $\mathbf{F}_\theta$ must be compensated by adding a vertical field which adds to B_θ over $3\pi/2<\theta<\pi/2$ and subtracts from B_θ over $\pi/2<\theta<3\pi/2$, as indicated in Figure 4.13.

The vertical field interacts with the plasma current I_ϕ (which is out of the page) to produce an inward force.

$$\mathbf{F}_v = \mathbf{I}_\phi \times \mathbf{B}_v$$

which just balances $\mathbf{F}_\theta$. The z component of the required vertical field is

$$B_{vz} \simeq \frac{\mu_0 I_\phi}{4\pi R_0}\left[\ln\left(\frac{8R_0}{a}\right) + \frac{p}{B_\theta^2/2\mu_0} - \frac{3}{2} + \frac{1}{2}l_p\right]$$

where l_p is the plasma internal inductance. (Inductance is defined in Chapter 8).

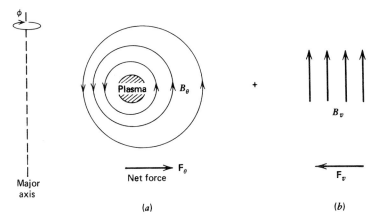

Figure 4.13 Force balance with a vertical field.

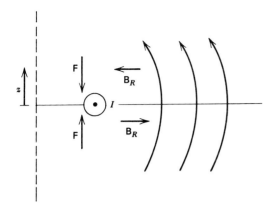

Figure 4.14 Restoring force against vertical displacement.

The radial component of the vertical field must be configured so as to interact with the plasma current to provide a restoring force

$$\mathbf{F} = \mathbf{I}_\phi \times \mathbf{B}_{vR}$$

if the plasma is vertically displaced from the horizontal midplane. A concave field, as shown in Figure 4.14, is required.

The requirement on vertical field curvature can be written

$$n \equiv -\frac{R}{B_v}\frac{\partial B_v}{\partial R} > 0$$

4.2.2
Stability

A plasma column which is confined entirely by an axial current I and associated field

$$B_0(r_p) = \frac{\mu_0 I}{2\pi r_p}$$

is unstable against pinch-type deformations of the type illustrated in Figure 4.15. Where the column pinches in the magnetic field, the inward magnetic pressure $B_0^2/2\mu_0$ is greater than the outward kinetic pressure; where the column bulges out, the inward magnetic pressure is weaker than the outward kinetic pressure. Thus, the deformation leads to force imbalances which increase the deformation.

The pinch instability can be suppressed by an axial magnetic field B_ϕ imbedded within the plasma. Associated with B_ϕ is a pressure $B_\phi^2/2\mu_0$ acting normal to the field. Since the toroidal field moves with the plasma (in the perfectly conducting limit) to conserve magnetic flux, $\Phi = \pi r_p^2 B_\phi = $ constant, the outward magnetic pressure $B_\phi^2/2\mu_0$ increases where the column pinches and decreases where the column bulges, thus providing a restoring force to oppose the deformation.

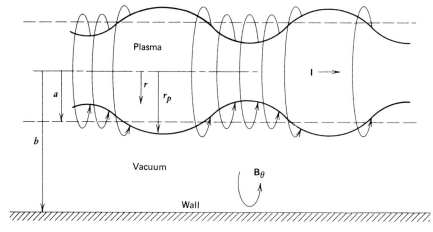

Figure 4.15 Pinch instability.

Since the poloidal magnetic flux $B_\theta(\theta)dr\ dl(\theta)$ per unit length of the column depends only upon the axial current, a kinking of the column produces regions of weaker and stronger poloidal fields, as shown in Figure 4.16.

The instability mechanism is similar to that for the pinch instability; where B_θ is stronger there is a net inward force that further kinks the column. Again, an axial

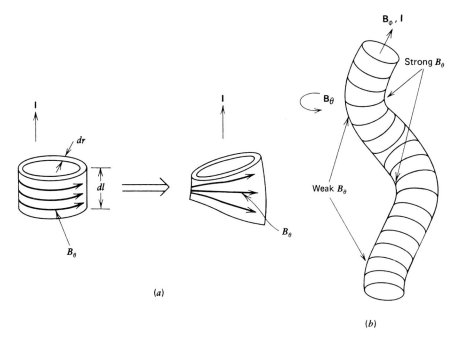

Figure 4.16 Kink instablity: (a) deformation of column unit length; (b) column deformation.

magnetic field imbedded in the plasma can suppress kink instabilities because the tension in the field lines resists the kinking.

For both the pinch and the kink instabilities the driving force is the poloidal field associated with the axial current, and the suppression mechanism is the axial field. Thus, it is plausible that these instabilities impose an upper limit on the allowable axial current for a given value of the axial magnetic field.

The same type of instabilities remain when a long plasma column is bent (figuratively) into a torus to form a tokamak plasma. The kink-mode stability limit for tokamaks can be characterized in terms of the safety factor

$$q(r) = \frac{B_\phi 2\pi r^2}{\mu_0 I(r) R_0} = \frac{r B_\phi^0}{R_0 B_\theta(r)}$$

where $I(r)$ is the amount of current carried in the region $0 < r' < r$ of the plasma. $q(r)$ is equal to the number of times a magnetic field line traces around the torus in the toroidal direction before it completes one revolution in θ through 2π.

Suppression of internal kink modes requires

$$q(0) \geq 1$$

and suppression of surface ($r=a$) kink modes requires

$$\frac{q(a)}{q(0)} \gtrsim 2-3$$

If the kinetic pressure is large in a region where the poloidal magnetic pressure is reduced because of a small radius of curvature of the plasma column, it is possible that a local bulging deformation, similar to that observed at a weak spot on a rubber inner tube, will occur. This type of "ballooning" instability imposes the limit

$$\frac{a}{R_0} \frac{p}{B_\phi^2(a)/2\mu_0} \equiv \frac{a}{R_0} \beta_\theta \lesssim 1$$

on tokamaks.

These stability limitations may be combined to obtain a limit on the parameter β_t which is important in determining the power density ($\sim \beta_t^2 B_\phi^4$) and ultimately the practicality of a tokamak,

$$\beta_t \equiv \frac{p}{B_\phi^2/2\mu_0} = \left(\frac{a}{R_0}\beta_p\right)\left(\frac{a}{R_0}\right)\left[\frac{q(0)}{q(a)}\right]^2\left[\frac{1}{q(0)}\right]^2$$

where the previous definitions of β_p and q have been used. The first factor is limited to values less than unity by ballooning modes. The second factor is limited, in future reactors, to values less than about one-third by geometry, shielding, and maintainability requirements. The third factor is limited to some value in the range of $\sim 0.1–0.3$ by surface kink modes. The last factor is limited to less than unity by internal kink modes. Theoretically the β-limit increases with plasma elongation (height/width).

4.2.3
Density Limit

More than a decade ago, a simple fit (now known as the Greenwald density limit)

$$n_G = I/\pi a^2$$

(n in $10^{20}/m^3$, I in MA, a in m) was found to bound the densities of stable operating points in three ohmic heated tokamaks, as shown in Figure 4.17

Over the years, this simple empirical fit has been found to provide a reasonable bound (to within a factor of about 2) on the density limit for a variety of tokamaks with shaping, auxiliary heating, and so on. However, stable discharges have been achieved with densities up to approximately a factor of 2 greater than the empirical Greenwald limit.

4.2.4
MHD Instability Limits

Computer simulations have shown that the Troyon β limit

$$\beta_c(\%) = \beta_N \frac{I}{aB_\phi}$$

provides a good representation of the ballooning mode limit on plasma pressure. Theoretical arguments indicate that the constant β_N should be proportional to the plasma internal inductance ℓ_i, and the experimental results shown in Figure 4.18 are well bounded by $\beta_c(\%) = 4\ell_i/aB_\phi$

The present tokamak operating experience is characterized by the Troyon expression above with $\beta_N \approx 2.5$–3.0, except when there is access to the second stability regime, which allows higher values of β_N.

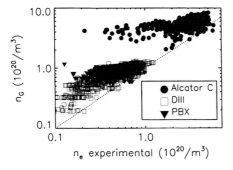

Figure 4.17 Comparison of Greenwald and experimental density limits (Greenwald, PPCF, IOP, London).

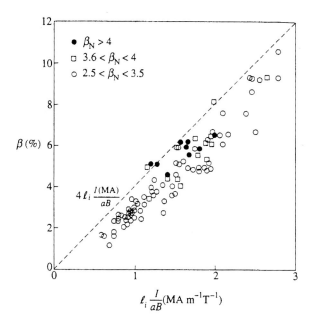

Figure 4.18 Experimental values of β plotted against $\ell_i I/aB$ for DIII-D (Taylor *et al.*, Proc. Thirteenth IAEA Conf. Plasma Phys, and Fusion Research, Vol. 1, p. 177, Vienna 1991).

As discussed in Section 4.2.2 above, surface kink modes impose a constraint $q(a)/q(0) \geq 2$–3 and internal kink, or interchange, modes impose a constraint $q(0) \geq 1$. Using $\ell_i = \ln(1.65 + 0.89(q_{\mathrm{edge}} - 1))$, the ($\ell_i$, q_{edge}) MHD stability diagram shown in Figure 4.19 can be constructed to display the combined limitation of the ballooning and kink modes. Here $q_{\mathrm{edge}} = q(a)$ for circular plasmas and $q_{\mathrm{edge}} = q_{95}$ for noncircular plasmas

$$q_{95} = \frac{5a^2 B}{2RI}(1 + \kappa^2)\left(1 + \frac{2}{3}\left(\frac{a}{R}\right)^2\right),$$

where the elongation $\kappa = \frac{b}{a}$ is the ratio of the vertical to horizontal radii of an elongated plasma.

The region of MHD stability is bounded to the left by the m=2 instability, which results in a disruption (a large-scale MHD instability that terminates the discharge) as $q_{\mathrm{edge}} \to 2$ from above. The central value of $q(0)$ is limited by sawtooth oscillations. The actual limiting value is about $q(0) \approx 0.7$, rather than the 1.0 that the simplified arguments suggest. This sawtooth (a high-frequency MHD instability) constraint limits the allowable values of ℓ_i and is most stringent at low q_{edge}. The resulting lines of constant β_c are shown in Figure 4.19. Again, access to the second stability regime will allow substantially higher values of β to be achieved.

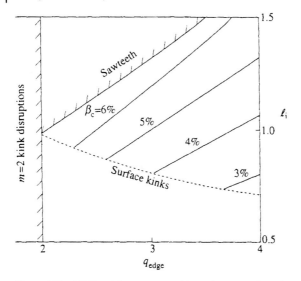

Figure 4.19 MHD stability operational boundaries in tokamaks (Wesson, Tokamaks, Oxford University Press, Oxford 1997).

4.2.5
Scaling of Experimental Energy Confinement Times

An "experimental" energy confinement time can be constructed from measured or otherwise known quantities

$$\frac{1}{\tau_E^{\text{exp}}} = \frac{P_{\text{HEAT}} - P_R - \dot{W}_{th}}{W_{th}}$$

where $W_{th} = \int nTdV$ is the total thermal energy of the plasma constructed by integrating the measured temperature and density profiles over the volume of the plasma, P_R is the total radiation from the plasma measured by bolometers, and P_{HEAT} is the known heating power input to the plasma.

One of the purposes of measuring energy confinement times is to check theoretical transport models. However, when theoretical and experimental energy confinement times are compared, they invariably differ. This has led to the widespread use of correlations of the measured experimental energy confinement times with machine and plasma parameters.

4.2.5.1 Ohmic Confinement
Early measurements on ohmically heated (no external heating) plasmas led to the correlation

$$\tau_E(s) = 0.07 \left(\frac{n}{10^{20}} \right) a R^2 q_{\text{edge}}$$

where n is the average electron density, a and R the minor and major plasma radii, and $q_{edge} = q_{95}$ is the cylindrical equivalent edge safety factor. (Note that the improvement in confinement with increased density is in conflict with the decrease in confinement predicted by neoclassical theory).

4.2.5.2 L-Mode Confinement

In order to improve the plasma energy content above that achieved with ohmic heating alone, additional heating was applied using beams of energetic neutral particles or RF waves. The results were initially disappointing in that, for given operational conditions, the confinement was found to degrade with increasing power. By regression analysis of the results from several tokamaks Goldston obtained the confinement scaling law that bears his name

$$\tau_G(s) = 0.037 \frac{I R^{1.75} \kappa^{0.5}}{P^{0.5} a^{0.37}}$$

(I in MA, P in MW), where P is the applied power. The degradation is apparent through the factor $P^{0.5}$ in the denominator.

Although the scaling was obtained before the large tokamaks such as JET were operational, it was found to describe the results on these machines as well. In order to improve the predictive capability for the proposed tokamak reactor ITER, an extended data base, including the larger tokamaks, was used to generate a more precise form of Goldston's scaling. The resulting confinement time, which was given the name ITER89-P, is given by

$$\tau_E^{ITER89-P}(s) = 0.048 \frac{I^{0.85} R^{1.2} a^{0.3} \kappa^{0.5} \left(n/10^{20}\right)^{0.1} B^{0.2} M^{0.5}}{P^{0.5}}$$

(I in MA, P in MW) where B is the toroidal magnetic field (T) and M is the mass in amu of the plasma ion species. A comparison between this formula and data from a number of tokamaks is shown in Figure 4.20.

4.2.5.3 H-Mode Confinement

It was found that when sufficient power was applied to an L-mode discharge that the discharge made an abrupt transition in which the edge transport was apparently reduced, leading to edge pedestals in the temperature and density. The effect of this was to produce roughly a doubling of the confinement time. The ability to achieve this high confinement (H-mode) condition has been observed to depend on the amount of non-radiative power flowing across the last closed flux surface, a threshold scaling relation for which is

$$P_{LH}(\text{MW}) = \left(\frac{2.84}{M}\right) n_{20}^{0.58} B^{0.82} R a^{0.81}$$

This behavior was subsequently obtained in many tokamaks and analogues of the ITER89-P L-mode scaling law have been derived for H-mode discharges. The results are represented by

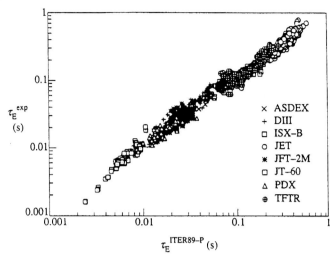

Figure 4.20 Comparison of the experimental values of confinement time from a number of tokamaks with the L-mode scaling τ_E^{ITER-P} (Yeshmanov *et al.*, Nucl. Fusion, 30, 1999 (1990)).

$$\tau_E^{ITER\,H93-P}(s) = 0.053 \frac{I^{1.06}\,R^{1.9}\,a^{-0.11}\,\kappa^{0.66}\,(n/10^{20})^{0.17}\,B^{0.32}\,M^{0.41}}{p^{0.67}}$$

(*I* in MA, *P* in MW). It is seen that the dependencies are generally similar to those of the L-mode.

A more recent and widely used H-mode correlation is

$$\tau_E^{IPB98(y.2)}(s) = \frac{0.056\,I^{0.93}\,B^{0.15}\,(n/10^{20})^{0.41}\,M^{0.19}\,A^{-0.58}\,R^{1.97}\,\kappa^{0.78}}{p^{0.69}}$$

where $A = \frac{R}{a}$, $P(MW)$ is the heating power, $I\,(MA)$ is the plasma current, and κ is the elongation.

The plasma confinement is usually observed to deteriorate significantly as the density approaches the Greenwald limit, as shown for the JET tokamak data in Figure 4.21. H97 = τ_E^{exp}/τ_{97}, where τ_{97} is an empirical H-mode confinement scaling relation similar to the ones given above.

4.3
Alternative Confinement Concepts

The tokamak concept is the main confinement concept worldwide in terms of both the state of advancement and the level of activity. However, there are a number of other concepts under active development. Three of these concepts – stellarator, spherical torus, and reversed field pinch – are discussed in this section.

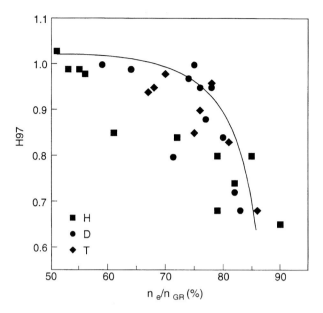

Figure 4.21 H-mode confinement degradation in JET as the Greenwald density limit is approached. (H97$=\tau_E^{exp}/\tau_{97}$) (Greenwald, PPCF, IOP, London).

4.3.1
Stellarator

The term stellarator is used to describe that class of toroidal confinement devices that produce closed flux surfaces entirely by means of external magnets (in contrast to the tokamak, in which a current in the plasma produces the poloidal magnetic field). The stellarator confinement concept was one of the first to be investigated. However, the success of the closely related tokamak in the late 1960s drew attention away from stellarator research.

The classical stellarator has ℓ pairs of helical conductors with alternately antiparalleled current flow that produces a poloidal field in the plasma. The net toroidal field in the plasma from these helical conductors is nulled, so a set of toroidal field coils is also required. The concept is illustrated in Figure 4.22.

The torsatron and heliotron configurations have ℓ helical conductors with parallel current flow. These helical windings also provide the toroidal field in the torsatron, while the heliotron requires a separate set of toroidal field coils. In general, a separate set of coils is required to provide a vertical field in the torsatron and heliotron.

The stellarator class of confinement concept offers two distinct potential advantages relative to the present tokamak. There is no inherent limitation to the length of operation because there is no continuous transformer action required, as is the case for the tokamak. Disruptive termination of the discharge, with the plasma energy being deposited on the chamber wall in an extremely short time – a worrisome problem with tokamaks – does not seem to occur in stellarators.

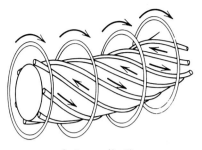

Stellarator (l = 3)

Torsatron (l = 3)

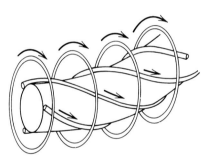

Heliotron (l = 3)

Figure 4.22 Stellarator, torsatron, and heliotron coil configurations. For the torsatron and heliotron, additional vertical-field windings (not shown) are required to maintain plasma equilibrium. (Reproduced with permission from R.L. Miller and R.A. Krakowski, Los Alamos Scientific Laboratory, Publication LA-8978-MS, 1981.)

Stellarators have a promising, but limited, experimental basis. Stellarator experiments are currently operating in the USA (CAT, HSX), Germany (W7-AS, WEGA), Japan (LHD, CHS), Spain (TJ1U, TJII, UST1) and Australia (H-1), the largest of which are the Large Helical Device in Japan and Wendelstein-7AS in Germany. A large experiment, Wendelstein-7X is under construction in Germany. A large USA stellarator experiment, NCSX, was recently canceled because of manufacturing difficulties.

4.3.2
Spherical Torus

The spherical torus (ST) is basically a tokamak with the central region reduced to the minimum size possible, leading to a smaller overall size. Theory indicates that plasma pressure can be confined in a ST with lower magnetic field strength than in a tokamak, providing the promise of higher values of β. There are several operational ST experiments, including NSTX, PEGASUS and HIT-II in the USA, MAST in the UK, GLOBUS-M in Russia, ETE in Brazil, TST-2 and LATE in Japan, and SUNIST in China. The National Spherical Torus Experiment (NSTX), which is depicted in Figure 4.23, has achieved $\beta \approx 30-40\%$ and has consistently achieved energy confinement times 2–3 times larger than predicted by conventional tokamak correlations. The ST lends itself to high "bootstrap" currents (current driven by the plasma pressure gradients, as opposed to transformer action), and bootstrap current fractions up to 50% have been achieved in NSTX.

4.3.3
Reversed-Field Pinch

The reversed-field-pinch (RFP) concept is similar to the tokamak in that it has a toroidal field produced by external coils and a poloidal field produced by an axial

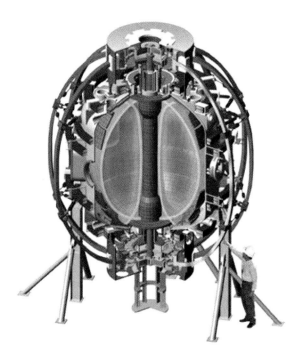

Figure 4.23 Schematic of the National Spherical Torus Experiment (NSTX) (Reproduced with permission from the NSTX website).

current induced in the plasma by external transformer coils. The unique feature of an RFP is the reversal of direction of the toroidal field in the outer region of the plasma. The field reversal should stabilize kink modes, thus relaxing the limitation on allowable plasma current, for a given toroidal field strength, found with tokamaks. This introduces the possibility of ohmically heating an RFP to thermonuclear temperature which, as will be discussed, is unlikely with tokamaks. The theoretical β limits are also higher for RFP, implying the possibility of achieving higher power densities than in a tokamak.

A comparison of the RFP and tokamak magnetic configurations is given in Figure 4.24.

The experimental data base for RFPs is extensive, based upon several experiments over a period of more than 25 yr. The establishment of the field reversed configuration has been convincingly demonstrated. However, the time constants, relaxation mechanisms, and energy requirements associated with establishing and maintaining the field reversed configuration are not well understood.

RFP experiments are currently operating in the USA (MST), Italy (RFX), Japan and Sweden (EXTRAP T2). A major recent development in RFP research was the control of MHD tearing modes, which increased the energy confinement time from 0.001 to

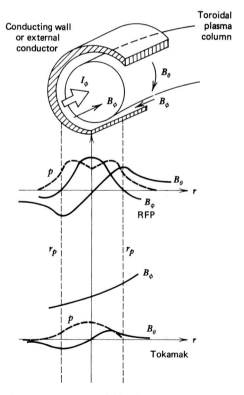

Figure 4.24 Magnetic field and pressure profiles for a reversed-field pinch (RFP) and a tokamak.

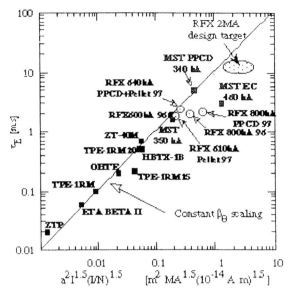

Figure 4.25 Confinement times achieved in RFP experiments (Reproduced with permission from the RFX website).

0.01 s, simultaneous with achievement of $\beta \approx 25\%$. The status of RFP research is best summarized by the achieved confinement times shown in Figure 4.25.

Problems

4.1. Estimate the fraction of ions in an initially isotropic velocity distribution that would escape immediately through the loss cone of a simple mirror with mirror ratio $R=2$.

4.2. Calculate the particle confinement time in a minimum-B mirror with $n_i=n_e=3\times10^{20}\mathrm{m}^{-3}$, $T_i=30\mathrm{keV}$, and mirror ratio $R=B_{max}/B_{min}=3$.

4.3. Calculate the relative potential ϕ_i between the end plug and the central solenoid for a tandem mirror (without thermal barrier) with plug electron density $n_p=5\times10^{20}\mathrm{m}^{-3}$, solenoid electron density $n_c=1\times10^{20}\mathrm{m}^{-3}$, and electron temperature $T_c=30\mathrm{keV}$.

4.4. Calculate the central solenoid particle confinement time for Problem 4.3.

4.5. If a thermal barrier with $n_b=5\times10^{19}\mathrm{m}^{-3}$ is introduced in Problem 4.3 and the plug electrons are heated to $T_p=75\mathrm{keV}$, what value of electron density in the plug n_p is now required to achieve the same confinement as in Problem 4.4?

4.6. Prove that in a purely toroidal field the $E\times B$ drift carries the plasma radially outward when B_ϕ is out of the page in Figure 4.8.

4.7. Compute the ∇B and curvature drift velocities in a tokamak with $a=1\mathrm{m}$, $R_0=4\mathrm{m}$, $B_\phi^0=7\mathrm{T}$, and $T_e=T_i=10\mathrm{keV}$.

4.8. Calculate the poloidal field at the outer surface ($r=a=1.5$m) of a tokamak plasma that carries 6MA current.

4.9. Estimate the vertical field strength, B_{vz} needed in a tokamak with $I=5$MA, $R_0=5$m, and $a=1$m operating with $n_i=n_e=2\times10^{20}$m^{-3} and $T_i=T_e=10$keV. (Neglect the plasma inductance.)

4.10. Calculate the value of the toroidal field B_ϕ^0 required to ensure stability against kink modes in the tokamak of Problem 4.9.

4.11. Use the L-mode empirical scaling law to estimate the energy confinement time in a tokamak with minor radius $a=1.5$m and density $n=2\times10^{20}$m^{-3}.

5
Plasma Heating

5.1
Ohmic Heating

For those confinement concepts, such as the tokamak and the RFP, for which a current flows in the plasma, collisional friction provides an intrinsic heating mechanism. The resistive, or ohmic, heating power deposition can be written

$$P_\Omega(\text{MW/m}^3) = 10^{-6}\eta j^2 = 2.8 \times 10^{-15} \frac{Z_{\text{eff}} I^2(A)}{a^4(\text{m}) T_e^{3/2}(\text{keV})}$$

where it has been assumed in the second equality that the current density j is uniformly distributed over the plasma cross section ($I = \pi a^2 j$), and the expression derived in Section 2.6 for plasma resistivity has been used. Z_{eff} is the effective charge averaged over all ions in the plasma.

Using Ampere's law,

$$\mu_0 I = 2\pi a B_\theta$$

and the definition of the safety factor,

$$q(a) \equiv \frac{a B_\phi}{R_0 B_\theta}$$

the ohmic heating power can be expressed as

$$P_\Omega(\text{MW/m}^3) = 7 \times 10^{-2} \frac{Z_{\text{eff}}}{T_e^{3/2}(\text{keV})} \left[\frac{1}{q(a)}\right]^2 \left(\frac{B_\phi}{R_0}\right)^2$$

This form illustrates two important points: (1) P_Ω saturates with increasing temperature, and (2) P_Ω is maximum in small high-field tokamaks or RFPs.

Now consider the possibility of heating a tokamak plasma to thermonuclear temperatures with ohmic heating. In order for the temperature to increase, the ohmic heating power must exceed the power loss due to heat conduction and convection and to radiation. Neglecting radiative losses,

Fusion: Second, Completely Revised and Enlarged edition. Weston M. Stacey
Copyright © 2010 WILEY-VCH Verlag GmbH & Co. KGaA, Weinheim
ISBN: 978-3-527-40967-9

$$P_\Omega > P_{tr} = 10^{-6} \frac{\frac{3}{2} nkT}{\tau_E}$$

Using the empirical expression for energy confinement in an ohmic heated plasma leads to an expression for the maximum temperature that can be achieved with ohmic heating,

$$T_e^{5/2}(\text{keV}) < 1.46 \frac{Z_{\text{eff}}}{[q(a)]^2} \left(\frac{aB_\phi}{R_0}\right)^2$$

For conventional tokamak parameters $[B_\phi \simeq 5\,\text{T}, R_0/a \simeq 3-5, q(a) \simeq 2-4]$ maximum temperatures on the order of ~ 1 keV are predicted. Thus, it does not appear likely that tokamaks can be heated to thermonuclear temperatures without some form of auxiliary heating. Ohmic heating to ignition would require much higher fields and/or smaller values of R_0/a; such an approach has been investigated.

5.2
Neutral Beam Injection (NBI)

Injection of highly energetic hydrogen or deuterium atoms into a plasma, where they become ions and give up their energy to the plasma ions and electrons via Coulomb scattering, has proven to be a highly successful means of plasma heating. The injected particles must be neutral in order to pass through the magnetic field surrounding the plasma.

A representative NBI system is depicted in Figure 5.1 for a deuterium beam. Ions are extracted from an ion source by an accelerating grid.

The principal creation process utilized in the sources consists of ionizing collisions between electrons and atoms or molecules. In an electron bombardment source, electrons of sufficient energy, coming from an electron source, are injected into the gas, and only these "primaries" are responsible for the ionization process. The liberated electrons are partly lost and partly accumulated, neutralizing the ion space charge, but they do not gain enough energy to participate in the ionization process. If the secondary electrons are heated (by electric fields or by collisions) so that they can materially contribute to the ionization rate, the process is an active electric discharge. With a suitable power input, these latter can give rise to highly ionized, dense plasmas.

Hydrogen atoms can attach a second electron, filling the $1s$ shell, to form stable negative ions with binding energies of 0.75 eV. Under suitable conditions, electric discharges in hydrogen yield surprising amounts of negative ions. D^- ion sources have produced 20 kA beams at 320 keV for 20s.

The more developed D^+ plasma ion sources produce positively charged atomic and molecular ions (e.g., D^+, D_2^+, and D_3^+). The accelerating grid provides the same momentum change to all ions, which results in the molecular ions having one-half and one-third the energy of the atomic ion species. The molecular species are sometimes separated out by passing the beam through a bending magnet.

Neutral Beam Injection (NBI)

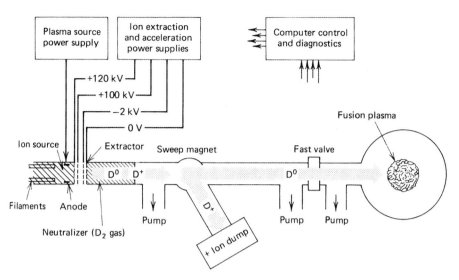

Figure 5.1   Schematic diagram of a beam line for a neutral injection system. (Reproduced by permission of E. Teller, ed., *Fusion*, Academic Press, 1981.)

The atomic ions are then passed through a neutralizer cell where they charge-exchange with deuterium gas, producing an energetic beam of neutral deuterons. The charge-exchange cross section decreases rapidly with increasing energy of the entering ion beam, so that the neutralization efficiency (i.e., the number of neutral deuterons produced per entering deuterium ion) decreases with beam energy, as shown in Figure 5.2. Also shown is the much better neutralization efficiency obtainable with negative deuterium ions, the technology for which is at an early stage of development.

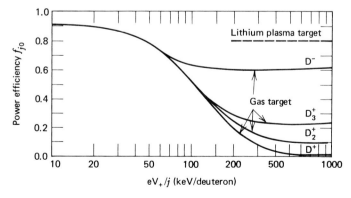

Figure 5.2   Maximum power efficiencies for the neutralization process $D^- \to D^0$ and $D_j^+ \to D^0, j = 1, 2, 3$, as function of energy per deuteron.

Remaining ions are removed from the beam by the sweep magnet. Otherwise the various reactor magnetic fields would bend the ions into surfaces near the entrance port, possibly releasing gas bursts or melting the surfaces. The considerable power in this ion beam must be handled by the ion-beam dump. The vacuum pumps distributed along the beam line remove most of the gas emerging from the neutralizer and the ion-beam dump and must maintain the pressure between the sweep magnet and the entrance port at a sufficiently low value so that very little of the neutral beam is reionized. Only an extremely small fraction of the beam particles can be allowed to strike any of the surfaces along the beam line and at the entrance port to the confinement region because gas evolution and reionization would otherwise lead to beam attenuation and to material damage.

The entire NBI system must be maintained under high vacuum, and the portion before the neutralizer cell must be shielded from stray magnetic fields. An NBI system designed for a tokamak ETR is depicted in Figure 5.3.

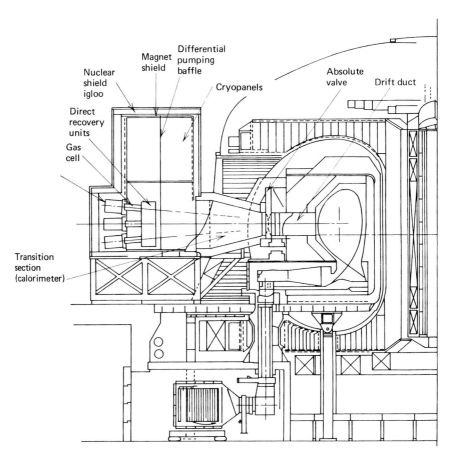

Figure 5.3 Elevation view of NBI system for a tokamak ETR. (Reproduced by permission of International Atomic Energy Agency, STI/PUB/619, 1982.)

Once the neutral particle beam enters the plasma, it is attenuated by charge-exchange and ionization reactions with the plasma ions and electrons, producing an energetic ion at the location of the reaction. This energetic ion moves along magnetic field lines on a flux surface, suffering numerous Coulomb scattering reactions with the plasma ions and electrons, in the process transferring its energy to those plasma ions and electrons on the same flux surface.

The attenuation of the neutral beam, hence the deposition of neutral beam energy, can be calculated from

$$\frac{dI_B}{ds} = -nI_B\langle\sigma v\rangle$$

where $n(\text{m}^{-3})$ is the plasma ion or electron density, $\langle\sigma v\rangle$ (m^3/sec) is the reactivity for charge-exchange and ionization, and $I_B(\text{m}^{-2}\text{sec}^{-1})$ is the flux of neutral particles. This equation has the solution

$$I_B(s) = I_B(0)e^{-n\langle\sigma v\rangle s} \equiv I_B(0)e^{-s/\lambda}$$

The mean free path λ for the attenuation of a beam of atomic mass A can be represented by the expression

$$\lambda(\text{m}) = \frac{5.5 \times 10^{17} E_B(\text{keV})}{A(\text{amu})n(\text{m}^{-3})Z_{\text{eff}}}$$

In order for the plasma to be effectively heated, most of the neutral beam energy must be deposited in the center of the plasma, rather than at the edge. Numerical simulations indicate that the plasma will be effectively heated if the mean free path is greater than about one-fourth of the plasma radius, that is, $\lambda \geq a/4$. Combining this result with the above expression for λ results in a criterion for the minimum energy of the neutral deuterons that is required for effective heating of a tokamak plasma,

$$E_B(\text{keV}) \geq \frac{n(\text{m}^{-3})a(\text{m})Z_{\text{eff}}A(\text{amu})}{2.2 \times 10^{18}}$$

The NBI power deposition density P_B must exceed the power loss rate in order for the plasma temperature to increase,

$$P_B > P_{\text{rad}} + P_{\text{tr}}$$

Using an expression for bremsstrahlung radiation which will be discussed subsequently and the empirical scaling law for τ_E, in tokamaks, this becomes

$$P_B(\text{MW}/\text{m}^3) > 4.8 \times 10^{-43}n^2(\text{m}^{-3})Z_{\text{eff}}T_e^{1/2}(\text{keV}) + 4.8 \times 10^{-2}\frac{T_e(\text{keV})}{a^2(\text{m})}$$

The total beam energy that must be input is equal to the amount required to offset losses (given approximately by the above expression integrated over the heating period) plus the energy required to raise n ions and n electrons in temperature by an amount ΔT, which energy is $3nk(\Delta T)$.

Table 5.1 Neutral beam parameters on tokamaks.

Device	E_B(keV)	P_{NB}(MW)
TFTR (USA)	120	40
DIII-D (USA)	80	20
JET (UK)	160	20
JT60-U (Japan, +ion)	100	40
JT60-U (Japan, −ion)	500	10

NBI heating has been used to achieve record plasma results, as indicated in Table 5.1. The present technology, based on positive ion sources, is being extended to D^- technology to meet the needs of future tokamak reactors. However, the NBI system is large and expensive and complicates the reactor design. Thus, there is an incentive to find another heating scheme. At present the most attention is being devoted to heating with radio-frequency waves.

5.3
Wave Heating

Collective wavelike motion can be established in the plasma (under the proper conditions) by launching electromagnetic waves into the plasma. Dissipation of the kinetic energy associated with the plasma wave motion via collisions can result in heating of the plasma.

5.3.1
Wave Phenomena in Plasmas

The fundamental properties of waves in plasma can be understood by considering low-frequency wave phenomena in a cold, uniform, uniformly magnetized plasma which has an equilibrium solution characterized by $j_0 = E_0 = v_0 = 0$ and $B_0 = $ constant. The governing equations for perturbed solutions are Ampere's and Faraday's laws,

$$\nabla \times \mathbf{B} = \mu_0 \mathbf{j} + \frac{1}{c^2} \frac{\partial \mathbf{E}}{\partial t}$$

and

$$\nabla \times \mathbf{E} = -\frac{\partial \mathbf{B}}{\partial t}$$

the momentum balance equations for ions and electrons,

$$n_i m_i \frac{\partial \mathbf{v}_i}{\partial t} = n_i e(\mathbf{E} + \mathbf{v}_i \times \mathbf{B})$$

$$n_e m_e \frac{\partial \mathbf{v}_e}{\partial t} = -n_e e(\mathbf{E} + \mathbf{v}_e \times \mathbf{B})$$

and the definition of current density,

$$\mathbf{j} \equiv e(n_i\mathbf{v}_i - n_e\mathbf{v}_e)$$

These equations may be linearized by expanding $\mathbf{B}$, $\mathbf{E}$, $\mathbf{v}_i$, and $\mathbf{v}_e$ about the equilibrium solution in the form $\mathbf{x}(r, t) = \mathbf{x}_0 + \mathbf{x}_1(r, t)$ and retaining only terms linear in the perturbations.

Now we seek the conditions under which wavelike solutions can be supported by writing

$$\mathbf{x}_1(\mathbf{r}, t) = \hat{\mathbf{x}}_1 e^{i(\mathbf{k}\cdot\mathbf{r} - \omega t)}$$

where $\mathbf{k}$ is the propagation vector and ω is the frequency of a traveling wave. Limiting consideration to waves which propagate in the direction of the equilibrium magnetic field $(\mathbf{k}\|\mathbf{B}_0)$, this waveform of the solution can be substituted into the linearized equations. Choosing $\mathbf{B}_0$ in the z direction, the resulting equations can be combined to obtain the component equations for any one of the perturbed variables, e.g. the electric field,

$$x : \left(c^2 k^2 - \omega^2 + \frac{\omega_{pi}^2 \omega^2}{\omega^2 - \Omega_i^2} + \frac{\omega_{pe}^2 \omega^2}{\omega^2 - \Omega_e^2}\right)\hat{E}_{1x} + i\left(\frac{\omega_{pi}^2 \Omega_i \omega}{\omega^2 - \Omega_i^2} + \frac{\omega_{pe}^2 \Omega_e \omega}{\omega^2 - \Omega_e^2}\right)\hat{E}_{1y} = 0$$

$$y : -i\left(\frac{\omega_{pi}^2 \Omega_i \omega}{\omega^2 - \Omega_i^2} + \frac{\omega_{pe}^2 \Omega_e \omega}{\omega^2 - \Omega_e^2}\right)\hat{E}_{1x} + \left(c^2 k^2 - \omega^2 + \frac{\omega_{pi}^2 \omega^2}{\omega^2 - \Omega_i^2} + \frac{\omega_{pe}^2 \omega^2}{\omega^2 - \Omega_e^2}\right)\hat{E}_{1y} = 0$$

$$z : \left(-\omega^2 + \omega_{pi}^2 + \omega_{pe}^2\right)\hat{E}_{1z} = 0$$

where

$$\omega_p^2 \equiv \frac{ne^2}{\varepsilon_0 m}, \quad \Omega^2 = \left(\frac{e B_{z0}}{m}\right)^2$$

are the plasma frequencies and gyrofrequencies for ions (i) and electrons (e).

One solution to this set of equations is

$$\omega^2 = \omega_{pe}^2 + \omega_{pi}^2 \simeq \omega_{pe}^2$$

which describes a wave propagating along the field (z) direction at the electron plasma frequency, without any constraint on allowable values of the propagation vector $\mathbf{k}$.

Other solutions occur when the determinant of the coefficients in the x- and y-component equations are required to vanish, which is the condition for the existence of a nontrivial solution. This condition may be written

$$c^2 k^2 - \omega^2 + \frac{\omega_{pi}^2 \omega}{\omega \pm \Omega_i} + \frac{\omega_{pe}^2 \omega}{\omega \pm \Omega_e} = 0$$

When $\omega \to |\Omega_i|$, or $\omega \to |\Omega_e|$, the propagation vector $k \to \infty$. This resonance condition implies that externally generated electromagnetic waves with frequencies $\omega \simeq |\Omega_i|$ or $\omega \simeq |\Omega_e|$ would readily drive plasma wave motion at these frequencies.

When the same procedure is carried through for waves that propagate in a direction perpendicular to the field ($\mathbf{k} \perp \mathbf{B}_0$), the condition for nontrivial solutions leads to $k \to \infty$ when

$$\omega^2 = \omega_{pe}^2 + \Omega_e^2$$

and when

$$\omega^2 = \frac{\omega_{pi}^2 \Omega_e^2}{\omega_{pe}^2 \Omega_e^2} \simeq |\Omega_e \Omega_i|$$

These two frequencies, known as upper hybrid and lower hybrid, respectively, also are resonance frequencies for excitation of plasma waves.

5.3.2
Waveguides

Electromagnetic waves can be transmitted within waveguides, which are long nonconducting cavities with conducting walls. Surface charges will arrange themselves within the conductor so as to terminate the normal component of the electric field, and the motion of these charges as the wave moves through the guide constitutes a current which terminates the tangential component of the magnetic field – consistent with Maxwell's equations,

$$\nabla \cdot \mathbf{D} = \rho$$
$$\nabla \times \mathbf{H} = \mathbf{j} + \frac{\partial \mathbf{D}}{\partial t}$$

Since the tangential component of the electric field and the normal component of the magnetic field are not terminated at the surface of a perfect conductor, the wave transmitted within a waveguide must have a resultant magnetic field which is transverse in the middle of the guide, but which bends and becomes axial at the sides of the guide. Two plane waves, of the same amplitude and frequency, traveling in a zigzag pattern, with multiple reflections from the wall, can achieve this condition if the plane waves cross at such an angle that they overlap exactly one wavelength in the width of the guide.

The general ideas discussed above are illustrated for a rectangular waveguide in Figure 5.4. Two plane waves are superimposed (Figure 5.4b) to form a traveling wave which has zero electric field at the location of the right and left vertical walls of the waveguide (Figure 5.4a). The resulting magnetic field is transverse in the center of the waveguide, but axial at the sides. The resultant wave is shown in Figure 5.4c. It propagates down the waveguide with a group velocity v_g, which is less than the propagation velocity $v_e = (\mu\varepsilon)^{-1/2}$ of the component waves,

$$v_g = v_e \sqrt{1 - \left(\frac{\lambda_e}{2b}\right)^2}$$

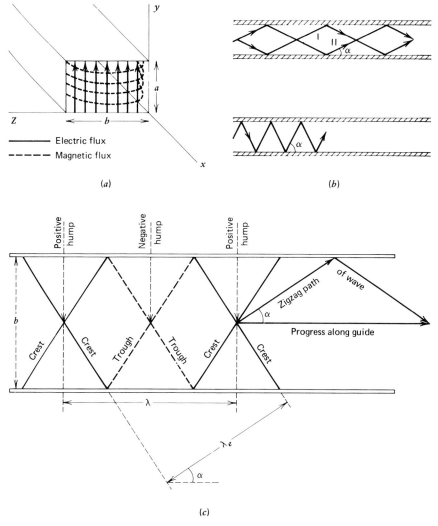

Figure 5.4 Wave transmission in a waveguide: (a) resultant fields; (b) superimposed waves; (c) resultant wave. (Reproduced by permission from *Fund. EM Waves*, Wiley, 1948.)

Here μ and ε are the permeability and the permittivity of the medium in the waveguide.

If the half-wavelength $\lambda/2$ of the resultant wave becomes as great as the width b of the waveguide, the two component waves will be reflected back and forth across the waveguide, producing a standing wave, but there will be no propagation of the wave along the guide. Thus, there is a cut-off wavelength λ_0,

$$\lambda_0 = 2b$$

and a corresponding cut-off frequency,

$$f_0 = \frac{v_e}{2b} = \frac{1}{2b\sqrt{\mu\varepsilon}}$$

Component waves with $\lambda_e > \lambda_0$ and $f_e < f_0$ will not be transmitted.

The discussion up to this point has been concerned with the propagation of the "dominant" mode. However, a waveguide can carry a variety of modes. The previous results can be generalized to modes that have m half-cycles in the y direction and n half-cycles in the z direction in Figure 5.4a. The group velocity and the wavelength of the resultant wave are

$$v_g = \frac{G}{\sqrt{\mu\varepsilon}} \quad \lambda = \frac{1}{\sqrt{\mu\varepsilon}\, f\, G}$$

where

$$G = \sqrt{1 - \frac{1}{4f^2\mu\varepsilon}\left(\frac{m^2}{a^2} + \frac{n^2}{b^2}\right)}$$

and the cut-off frequency is determined from the condition $G(f_0) = 0$,

$$f_0 = \sqrt{\frac{1}{4\mu\varepsilon}\left(\frac{m^2}{a^2} + \frac{n^2}{b^2}\right)}$$

A cylindrical waveguide can also be employed. In particular, a coaxial waveguide with an inner and outer conductor, as shown in Figure 5.5, is frequently used. The cutoff wavelength for the mode with the lowest cutoff frequency in a coaxial waveguide is

$$\lambda_0 \simeq \frac{2\pi(r_2 + r_1)}{2}$$

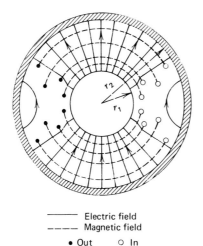

——— Electric field
– – – Magnetic field

• Out   ○ In

Figure 5.5 Coaxial waveguide. (Reproduced by permission from *Fund. EM Waves*, Wiley, 1948.)

A wave is launched into a waveguide by generating an electromagnetic field that resembles the field of the desired mode. Various types of resonating power sources may be used for this purpose. The electromagnetic field launched into the waveguide is usually a complex combination of many modes. The waveguide will then be designed so that only the desired mode will propagate, the others being beyond cutoff.

The transmitted power is given in terms of the electric and magnetic fields by the Poynting vector

$$\mathbf{P} = \mathbf{E} \times \mathbf{H}$$

5.3.3
Launching an Electromagnetic Wave into a Plasma

The externally generated electromagnetic (EM) wave is transmitted through the shield (and blanket) region to the plasma surface in a waveguide. If the end of the waveguide is simply left open, energy will radiate into the plasma. However, radiation from an abruptly terminated waveguide is generally not very efficient, because the EM fields in the waveguide usually cannot pass directly into being fields in the plasma region – there is a mismatch, and energy is reflected back into the guide. If the end of the guide is flared into a horn to provide a gradual transition from the modes of the waveguide to those of the plasma region, the radiation from the horn may become quite effective.

It becomes increasingly difficult to launch EM waves directly from a waveguide (even with a horn) as the wavelength increases: an antenna structure is used in these cases to launch the wave. The basic ideas involved can be illustrated by considering a straight antenna which has a length equal to one-half the wavelength. The transmitted EM wave is used to drive a current in the antenna. The resulting electric and magnetic fields at a distance r from the center of the antenna (Figure 5.6) are

$$E_\theta = \sqrt{\frac{\mu_0}{\varepsilon_0}} \frac{I_0}{2\pi r} \cos \omega \left(t - \frac{r}{c} \right) \frac{\cos[(\pi/2)\cos \theta]}{\sin \theta}$$

and

$$H_\phi = \frac{I_0}{2\pi r} \cos \omega \left(t - \frac{r}{c} \right) \frac{\cos[(\pi/2)\cos \theta]}{\sin \theta}$$

where I_0 is the maximum current in the antenna, c is the speed of the wave in the plasma (speed of light), and $\omega = 2\pi f$, f being the frequency of the transmitted wave in Hertz. The power launched into the plasma would be $P = E_\theta H_\phi$. More general relations can be derived for curved antennas of arbitrary length.

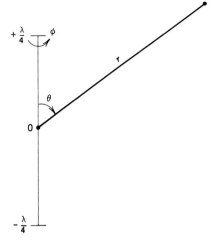

Figure 5.6 A simple half-wave antenna.

5.3.4
Radio-Frequency Heating

A substantial amount of research has been devoted to the use of externally generated electromagnetic waves to heat plasmas. This mode of heating is usually referred to as radio-frequency (rf) heating, although microwave frequencies are included in the spectrum of interest. The three frequency ranges of the greatest interest correspond to the electron gyrofrequency [electron cyclotron (ECH)], the lower hybrid frequency (LH), and the ion gyrofrequency [ion cyclotron range of frequencies (ICRF)].

An rf heating system consists of a power source or oscillator, a transmission system between the power source and plasma, and a launcher which directs the external electromagnetic wave energy into the plasma. Characteristics of rf heating systems are summarized in Table 5.2. The power sources that exist today are adequate, with minor upgrading, for reactor applications in the case of ICRF and LH systems. However, the gyrotrons that will be required for reactor application of ECH represent a major technological development. The transmission technologies needed for future reactors essentially exist today.

The launching structures presently used are illustrated in Figure 5.7. For the higher frequency (shorter wavelength) waves, waveguides are used, but for ICRF all experiments to date have used an antenna. The development of an antenna which can survive in a reactor plasma environment or the development of waveguide launching techniques represent developments required to extrapolate ICRF launching technology to reactors. The waveguide launching technology presently available for ECH and LH should be satisfactory (with some extensions) for reactors.

Once the electromagnetic wave is launched into the plasma, it is either reflected or it propagates into the plasma until it reaches a location where the plasma density and the magnetic field strength are such that one of the plasma resonance frequencies

Table 5.2 Characteristics of rf heating systems. Reproduced with permission from W.M. Stacey *et al.*, Georgia Institute of Technology report GTFR-38, 1983.

Heating Mode	*f* (GHz)	Power Source	Transmission	Launch		
				Current Practice	Future Practice	Wavelength
Electron cyclotron (ECH)	100–200	Gyrotron	Coaxial	Waveguides	Waveguides	1 mm
Lower hybrid (LH)	1–8	Klystron	Waveguides	Phased Waveguides	Phased Waveguides	10 cm
Ion cyclotron (ICRF)	0.025–0.1	Tetrode/triode	Waveguides	Antenna structure	Waveguides	10 m

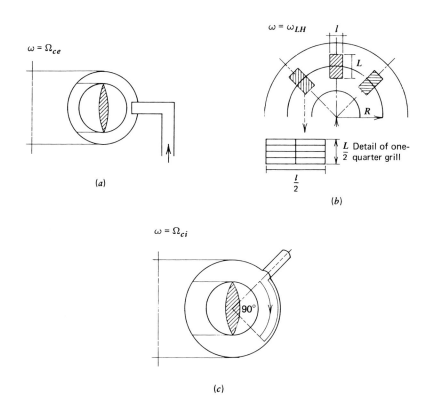

(a)

(b)

(c)

Figure 5.7 Launching structures: (a) oversize waveguides; (b) phased waveguide; (c) loop antenna. (Reproduced with permission of IAEA from STI/PUB/556, 1980.)

Table 5.3 RF and microwave heating powers on tokamaks (MW).

Device	ICRH	LHR	ECH
TFTR (USA)	11.0		
DIII-D (USA)	4.4		2.0
JET (UK)	20.0	7.0	
JT60-U (Japan)	7.0	8.0	
K-STAR (Korea)	6.0	3.0	4.0

corresponds to the wave frequency, at which point the energy of the external wave is transferred totally or partially to the plasma to set up waves in the plasma. (The shaded regions in Figures 5.7*a* and *c* correspond to such resonance regions in a typical plasma.) The energy of the plasma waves is ultimately dissipated by collisions among the particles, thereby heating the plasma.

Plasma RF heating experiments have met with mixed success. ICRF and ECH experience has been good, but LHR experience has been mixed. On the other hand, LHR experience for driving current has been good. Table 5.3 displays the heating power on existing tokamaks.

At the moment, ICRF heating seems the most promising candidate for extrapolation to reactor-grade plasmas. The major developmental requirement is a launcher that can survive in the reactor plasma environment.

5.4
Compression

An increase in magnetic field strength will, in general, compress and heat a plasma. If the compression occurs over a sufficiently long period of time, the law of adiabatic compression,

$$pV^{\gamma} = \text{constant}$$

is obeyed. The compression velocity must be small compared to the thermal velocity in order for the compression to be adiabatic. (A compression taking place on a much faster, microsecond, time scale is known as an implosion and is governed by a different set of laws.) The gas constant $\gamma = (2 + \delta)/\delta$, where δ is the number of degrees of freedom of the compression. If the compression is physically in one or two dimensions and the compression time is short compared to the scattering time $(\tau_{90})_L$ for large-angle deflection, then $\delta = 1$ or 2. If the compression time is comparable to or greater than $(\tau_{90})_L$, then the energy increase produced by the compression is shared among all three directions due to scattering deflections, independent of the dimensionality of the physical compression. Because the collision time for electrons is $\sim \sqrt{m_e/m_i}$ times the collision time for ions, it is possible to have a compression with $\delta = 3$ for electrons and $\delta < 3$ for ions.

Noting that the specific volume V is proportional to the inverse of the particle density n^{-1}, and that the pressure p is related to the average energy per particle W by $p = nW$, we can rewrite the above equation as

$$n^{1-\gamma} W = \text{constant}$$

When the compression time is less than $(\tau_{90})_L$, only the energy component, $W_\parallel$ in the direction of the compression changes. Denoting initial and final states by superscripts 1 and 2, respectively, this equation yields

$$\frac{W_\parallel^{(1)}}{W_\parallel^{(2)}} = \left[\frac{n^{(2)}}{n^{(1)}}\right]^{1-\gamma}$$

For an initially isotropic distribution, $W_\parallel^{(1)} = \delta W^{(1)}/3$ and $W_\perp^{(1)} = (3-\delta) W^{(1)}/3$, where δ is the number of directions of compression. Noting that $W_\perp^{(2)} = W_\perp^{(1)}$, we find

$$\frac{W^{(2)}}{W^{(1)}} \equiv \frac{W_\parallel^{(2)} + W_\perp^{(2)}}{W_\perp^{(1)} + W_\parallel^{(1)}} = \frac{3-\delta + \delta[n^{(2)}/n^{(1)}]^{-1}}{3}$$

We see that, for a given density compression $n^{(2)}/n^{(1)}$, a one-dimensional compression is more effective than a two-dimensional compression, which in turn is more effective than a three-dimensional compression in raising the particle energy.

5.5
Fusion Alpha Heating

Once the plasma temperature is raised by other heating methods to the point at which a significant amount of fusion takes place, the 3.5-MeV alpha particles produced in D-T fusion events constitute a self-heating mechanism. The charged alpha particles will follow magnetic field lines and will remain in the plasma to give up their energy collisionally to the plasma ions and electrons, unless their excursions from the flux surface due to curvature and ∇B drifts and to trapping in magnetic wells are sufficiently large to cause the alpha particles to strike the wall. In a tokamak these excursions are inversely proportional to the local poloidal field. Since the poloidal field is proportional to the enclosed plasma current by Ampere's law, alpha particle confinement and alpha heating both improve with increasing plasma current. Several megaamperes of plasma current is sufficient to confine most of the alpha particles in a tokamak.

5.6
Non-Inductive Current Drive

In the "conventional" tokamak concept the plasma current is driven inductively by the transformer action of the poloidal field coil systems, particularly the central

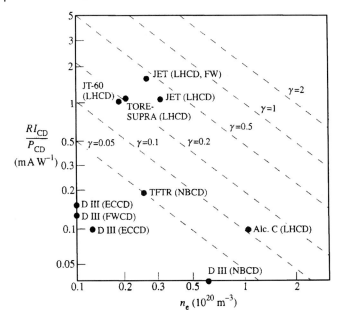

Figure 5.8 Experimentally achieved current-drive efficiencies.

solenoid coil. In order to maintain the plasma current against resistive losses it is necessary to continue to increase the field in the coils until the limiting field is reached and then to shut down the plasma and repeat the process. This operating scenario produces pulsed electromagnetic forces and thermal stresses, which in turn produce fatigue and the growth of micro-cracks.

A major plasma physics research effort is being devoted to finding ways to drive the plasma current in tokamaks non-inductively to achieve quasi-steady state operation. The basic idea is to impart toroidal momentum to plasma ions or electrons or to alter the rate at which these particle lose momentum, thereby creating a toroidal current. This can be done with tangentially injected neutral beams and with radio-frequency or microwave electromagnetic wave injection into the plasma. Experimentally achieved current-drive efficiencies are shown in Figure 5.8 (the quantity $\gamma = RIn_e/P_{cd}$).

5.7
ITER Plasma Heating and Current Drive Systems

The plasma heating and non-inductive current drive system specified for ITER is described in Table 5.4. A neutral beam injection system is specified to provide both heating and non-inductive current drive. An ion cyclotron resonance (ICRH) radio-frequency heating system is specified for additional or alternative plasma heating. A

Table 5.4 ITER plasma heating system powers (MW).

Neutral Beams (D⁻)	ICRH	LHR	ECRH
50	40	40	20

lower hybrid resonance (LHR) non-inductive current drive system is specified. An electron cyclotron resonance heating (ECRH) system is specified for plasma control.

The ITER negative ion source is required to produce a 40 A ($200 \, \text{A/m}^2$) current of D^- ions, which a vacuum insulated accelerator is then required to accelerate to 1 MeV in order to penetrate the large and dense ITER plasma. To date (2008), H^- ion beams of $146 \, \text{A/m}^2$ have been produced and accelerated to 836 keV; thus the achieved beam energy and current density are approaching the ITER requirements.

The ECRH power source consists of sets of 170 GHz, 2 MW gyrotrons.

Problems

5.1. Consider a tokamak plasma with $R_0 = 5 \, \text{m}$, $a = 1.2 \, \text{m}$, $Z_{eff} = 1.5$, and $q(a) = 2.5$. Calculate the maximum temperature achievable by ohmic heating for $B_\phi = 4$, 8, and 12 T.

5.2. Assuming $Z_{eff} = 1$ and $q(a) = 2$, calculate the value of aB_ϕ/R_0 required in order to be able to heat a tokamak plasma ohmically to thermonuclear temperatures ($\sim 7 \, \text{keV}$).

5.3. Consider a tokamak with $R_0 = 5.2 \, \text{m}$, $a = 1.2 \, \text{m}$, $B_\phi = 5.5 \, \text{T}$, $Z_{eff} = 1.2$, and $n = 1 \times 10^{20} \, \text{m}^{-3}$. Calculate the deuteron energy E_B required for effective neutral beam heating of the plasma.

5.4. Calculate the beam power deposition density P_B and the total beam power input required to heat the plasma from 1 to 7 keV for Problem 5.3. Calculate the total beam energy deposited in the plasma during the heating.

5.5. Calculate the frequencies required for ECH, LH, and ICRF heating of a deuterium plasma with density $n = 2 \times 10^{20} \, \text{m}^{-3}$ and magnetic field $B_{z0} = 4.5 \, \text{T}$.

5.6. Calculate the wavelengths corresponding to the frequencies of Problem 5.5.

5.7. Consider adiabatic compression in a tokamak in which the toroidal and poloidal magnetic fluxes are conserved (i.e., $a^2 B_\phi =$ constant and $q(a) =$ constant). Work out the scaling with major radius R and minor radius a under compression of the plasma density n the toroidal beta β_t, and the plasma temperature T.

6
Plasma–Wall Interaction

6.1
Surface Erosion

The first material surface facing the plasma has incident upon it a variety of particle and energy fluxes. Since the magnetic confinement is not perfect, there are fluxes of charged particles – electrons, plasma ions, alpha particles from the fusion reaction, impurity ions that have been previously eroded from the surface, and ions from the NBI process. Most of the plasma ions recycle from the wall as "cool" neutral atoms, and some of these undergo charge-exchange with plasma ions, creating thereby a flux of energetic "charge-exchange neutrals" incident upon the surface. The magnitude of the charged particle flux can be estimated from

$$\Gamma_w(\#/\mathrm{m}^2 \cdot \mathrm{sec}) = \frac{\bar{n}(\mathrm{m}^{-3})}{\tau_p(\mathrm{sec})} \frac{\mathrm{volume}(\mathrm{m}^3)}{\mathrm{area}(\mathrm{m}^2)}$$

where $\bar{n}$ is the average density, τ_p is the particle confinement time, volume refers to the plasma, and area refers to the wall.

An energetic ion or neutral incident on a material surface will transfer its energy to the lattice atoms, by scattering collisions, within the first micrometer or so. The displaced lattice atoms will have subsequent collisions with other lattice atoms. If, as a result of this energy transfer process, an atom on the surface receives an amount of energy in excess of its surface binding energy, then that atom will be ejected from the surface. This process is known as physical sputtering. The physical sputtering yields (atoms ejected per incident particle) for deuterons incident upon a number of different materials are given in Figure 6.1. Theoretically the physical sputtering yield should scale directly with the mass of the incident particle, so these curves can be scaled by $\frac{1}{2}, \frac{3}{2}$, and 2 to estimate sputtering yields for protons, tritons, and alpha particles, respectively. Kinematic considerations indicate that there is a threshold value E_t of the incident particle energy below which the energy transferred is less than the surface binding energy U_0 and no sputtering occurs. This threshold energy depends upon the masses of the incident (1) and surface (2) atoms,

$$E_{t1} = \frac{(m_1 + m_2)^2}{4 m_1 m_2} U_{02}$$

Fusion: Second, Completely Revised and Enlarged edition. Weston M. Stacey
Copyright © 2010 WILEY-VCH Verlag GmbH & Co. KGaA, Weinheim
ISBN: 978-3-527-40967-9

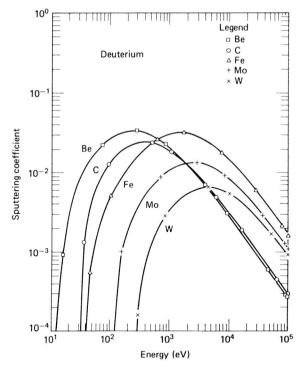

Figure 6.1 Physical Sputtering Yield for a Number of Materials for Deuterons. (Reproduced with permission from W. M. Stacey, *et al.* "FED-INTOR," Ga. Inst. Techn. report, 1982.)

For some surface materials, notably carbon, the incident hydrogen isotope apparently combines chemically with the surface material to form a compound with reduced surface binding energy. This chemical sputtering effect can enhance the sputtering yield by more than a factor of 2.

The atoms eroded from the surface enter the plasma, become ionized, and subsequently are incident upon the surface as energetic ions. The self-sputtering yield for these reincident ions can be quite large, as shown in Figure 6.2. For the higher atomic number – high-z – materials, the self-sputtering yield can exceed unity for ion energies in excess of a few hundred electronvolts. This situation potentially could lead to an "avalanche" effect which would rapidly increase the concentration of eroded ions in the plasma above tolerable limits.

Under certain conditions, instabilities can cause an extremely rapid loss of confinement, which will cause the plasma thermal energy to be deposited on the material surface, leading to melting and vaporization. Calculated melting zone thicknesses for parameters typical of a tokamak reactor are shown in Figure 6.3. The corresponding evaporation thicknesses are shown in Figure 6.4.

Erosion of the first wall surface by sputtering and by evaporation can limit the lifetime of the first wall. This erosion also produces impurities in the plasma, which can have very deleterious effects upon the plasma, primarily due to radiative processes.

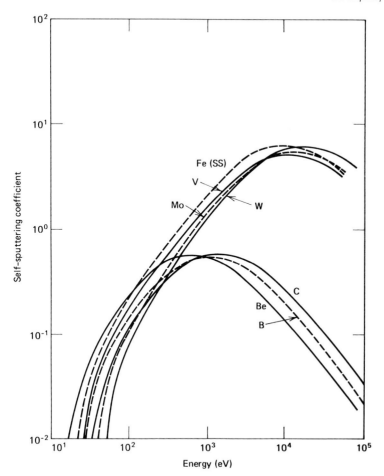

Figure 6.2 Self-Sputtering Yield for a number of materials. (Reproduced with permission from W.M. Stacey, *et al.* "FED-INTOR", Ga. Inst. Techn. report, 1982.)

6.2
Impurity Radiation

The predominant effect of impurities upon a plasma is radiation cooling. The impurity ions interact with the plasma electrons via several mechanisms to reduce the electron energy and produce electromagnetic radiation which escapes from the plasma.

When an electron is accelerated, there are electromagnetic fields created. The radiative power in these fields is given by

$$W_{\mathrm{rad}} = \frac{1}{6\pi} \frac{e^2 |\dot{\mathbf{v}}|^2}{\varepsilon_0 c^3}$$

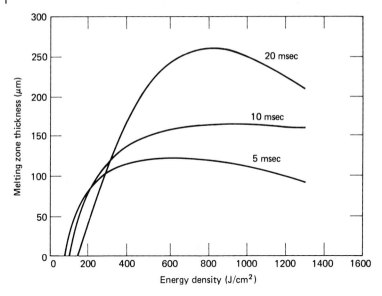

Figure 6.3 Stainless-steel melting zone thickness as a function of energy density and disruption time. (Reproduced with permission from C.C Baker *et al.* Demonstration Reactor Study, Argonne Nat Lab, 1983.)

The Coulomb interaction of an electron with an ion may be approximated as an interaction taking place over a time τ and a distance x. Thus, the electron is subjected to a force,

$$F = \frac{ze^2}{4\pi\varepsilon_0 x^2} = m_e\dot{v}$$

for a time

$$\tau \simeq \frac{x}{v}$$

so that the acceleration is

$$|\dot{\mathbf{v}}| \simeq \frac{ze^2}{4\pi\varepsilon_0 m_e x^2}$$

See Figure 6.5.

The total power loss per unit volume from this process, which is known as bremsstrahlung, is obtained by integrating over all such interactions,

$$P_{\mathrm{brems}} = n_z n_e |v_e| \int_{x_{\min}}^{x_{\max}} W_{\mathrm{rad}}\left(\frac{x}{|v_e|}\right) 2\pi x \, dx$$

which can be evaluated to obtain

$$P_{\mathrm{brems}}(\mathrm{MW/m^3}) \simeq 4.8 \times 10^{-43} z^2 n_z(\mathrm{m^{-3}}) n_e(\mathrm{m^{-3}}) T_e^{1/2}(\mathrm{keV})$$

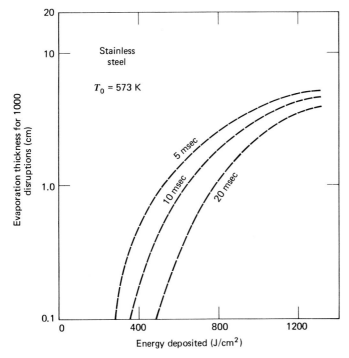

Figure 6.4 Evaporation thickness of stainless steel for 1000 disruptions for different Energy deposited. (Reproduced with permission from C. C. Baker *et al.* Demonstration Reactor Study, Argonne Nat Lab, 1983.)

This power is in the form of photons distributed roughly continuously in energy up to an energy about equal to the plasma electron mean thermal energy.

Collisional excitation of an orbital electron in a partially ionized impurity atom by a plasma electron causes the plasma electron to lose an amount of energy ΔE equal to the difference in energy levels of the ground and excited states. Normally the excited state will immediately decay to the ground state, accompanied by the emission of a photon with a discrete frequency v, where $hv = E$. The power loss due to this "line radiation" can be approximated by

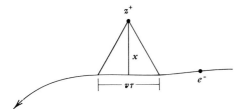

Figure 6.5 Impulse interaction approximation.

$$P_L(\text{MW/m}^3) \simeq 1.8 \times 10^{-44} \frac{z^4 n_e(\text{m}^{-3}) n_z(\text{m}^{-3})}{T_e^{1/2}(\text{keV})}$$

Free plasma electrons will recombine with partially ionized impurity atoms to form an ion in a charge state lower by 1. The energy loss to the plasma is approximately the plasma electron thermal energy, which is emitted in the form of a photon. This "recombination" radiation power loss can be approximated by

$$P_R(\text{MW/m}^3) \simeq 4.1 \times 10^{-46} \frac{z^6 n_e(\text{m}^{-3}) n_z(\text{m}^{-3})}{T_e^{3/2}(\text{keV})}$$

The radiative power loss from a plasma depends strongly upon the impurity concentration and atomic number z, and in a rather complicated manner upon the electron temperature. The normalized radiative power loss due to bremsstrahlung, line, and recombination radiation is shown in Figure 6.6 for several types of impurity ions. At very low plasma electron temperatures increasing T_e increases the ionization

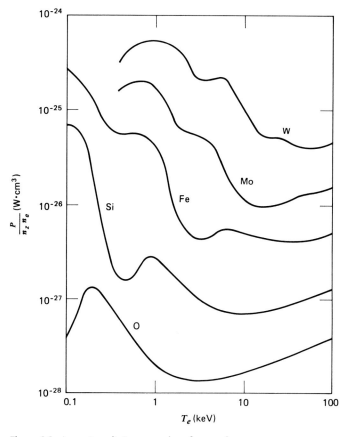

Figure 6.6 Impurity radiative power loss from a plasma.

state, hence the atomic radiation. At higher plasma temperatures the ion becomes fully stripped and the atomic radiation decreases with T_e. Finally, when the ion is fully ionized, there is only bremsstrahlung.

The effect of the impurity radiation on plasma performance can be evaluated by using the above expressions in the plasma power balance

$$P_\alpha = P_{tr} + P_{rad}$$

A crude approximation can be obtained by neglecting transport losses ($P_{tr} \rightarrow 0$), and using

$$P_\alpha = \tfrac{1}{4} n_i^2 \langle \sigma v \rangle U_\alpha$$

where n_i is the main plasma ion density and U_α is the 3.5-MeV fusion energy remaining in the plasma, results in an expression for the maximum allowable impurity concentration for which a self-sustaining fusion reaction is possible. This condition of the plasma power balance being satisfied without any external heating source is referred to as "ignition"

$$\frac{1}{4} \langle \sigma v \rangle U_\alpha \geq \frac{n_z}{n_i} W_{rad}(T_e, z)$$

where W_{rad} is obtained from the above expressions,

$$W_{rad} \equiv \frac{P_{brems} + P_L + P_R}{n_z n_e}$$

and is the quantity plotted in Figure 6.6.

This maximum permissible impurity concentration is plotted in Figure 6.7. It is apparent that only miniscule quantities of intermediate- and high-z impurities can be tolerated in a plasma. Thus, there is clearly a need for some means of preventing the impurity ions that are sputtered from the first wall surface from entering the plasma.

6.3
Impurity Control

Impurity control refers to those techniques that are employed to maintain the plasma purity within acceptable limits by preventing wall-eroded atoms from entering the plasma and/or by exhausting wall-eroded or fusion product ions from the plasma. This issue has been studied extensively for tokamaks.

Conceptually the simplest solution to the wall-eroded impurity problem is to prevent the erosion from taking place. Since it is practically impossible to eliminate the particle fluxes emanating from the plasma, prevention of erosion can only be accomplished by reducing the energy of these particles below the threshold energy for sputtering. This threshold energy varies from a few electronvolts for the low-z materials to a few hundred electronvolts for the high-z materials, as discussed in

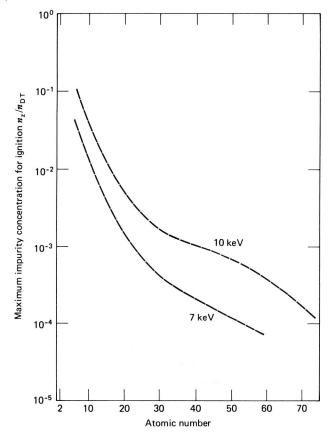

Figure 6.7 Maximum impurity concentration for which ignition can be achieved.

Section 6.1. Calculations and extrapolation of present experimental results indicate that the most probable energy of particles emanating from the edge of a tokamak reactor plasma will be in the range of ~100–300 eV. However, radiative cooling of the plasma edge by wall-eroded impurities may act to reduce the edge temperature without adversely affecting the plasma center if a medium-z impurity (e.g., stainless steel) which becomes entirely ionized at the central plasma temperatures is used. Whether such a "cold, radiating edge" condition will be stable remains an open question.

Insulating the plasma from the wall by interposing a region of high-density "cold" neutral gas has also been suggested as a means of reducing the energy of the particle fluxes striking the wall to below the sputtering threshold. Calculations indicate that the thickness of such a "gas blanket" would have to be so great (~1 m) as to make it impractical.

Perhaps the next simplest solution to the wall-eroded impurity problem is to mitigate the adverse effects of radiative cooling by using a low-z wall surface.

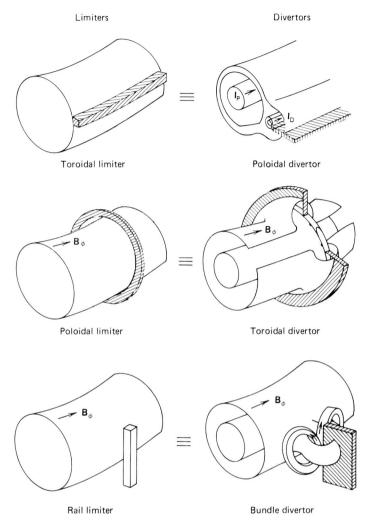

Limiters   Divertors

Toroidal limiter   Poloidal divertor

Poloidal limiter   Toroidal divertor

Rail limiter   Bundle divertor

Figure 6.8 Various limiter and divertor configurations.

Reference to Figure 6.7 shows that concentrations up to ~10% of low-z impurities may be tolerable in a plasma.

The technique that has been most successful to date involves magnetic diversion of the particle fluxes emanating from the plasma into a separate chamber. The principle, which is illustrated in Figure 6.8, is to divert some or all of the magnetic field lines near the walls of the plasma chamber out of the plasma chamber into a separate divertor chamber. The charged particles approaching the wall are then swept out of the plasma chamber into a divertor chamber where they are finally incident upon a divertor collector plate. A representative poloidal divertor configuration for a tokamak is shown in Figures 6.9 and 6.10. A fraction of the particles reflected from the divertor collector plate are pumped out a vacuum duct, and the remainder remain

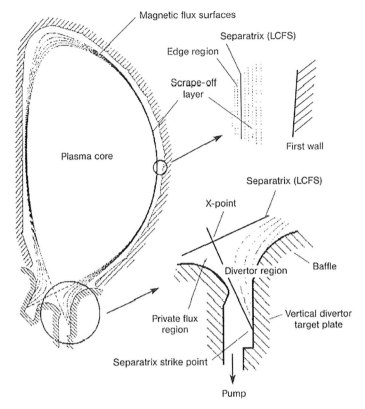

Magnetic flux surfaces

Separatrix (LCFS)

Edge region

Scrape-off layer

First wall

Plasma core

Separatrix (LCFS)

X-point

Baffle

Divertor region

Vertical divertor target plate

Private flux region

Separatrix strike point

Pump

Figure 6.9 A single-null poloidal divertor configuration.

in the divertor chamber as a cold partially ionized gas which serves to insulate the collector plate from the plasma by cooling the particle fluxes from the plasma as they pass through. Impurity atoms eroded from the collector plate are impeded from returning to the plasma chamber by this high-density gas and by the divertor action itself once they become ionized.

Divertor experiments have demonstrated conclusively that the impurity content of a plasma can be significantly reduced by magnetic divertors. For reasons that are not entirely understood, the magnetic configuration associated with the divertor also leads to improved energy confinement, altogether independent of the effect on plasma purity. The principal disadvantage of the divertor is that the required magnetic coil systems are complicated and expensive when extrapolated to a reactor.

An alternative to the divertor is to use a material "limiter" in the plasma chamber, as depicted in Figure 6.8. Particles reaching the edge of the plasma chamber strike the limiter, where they are reflected as neutrals and also cause sputtering of the limiter. If a vacuum pumping duct is located so that a significant fraction of the neutral particles reflected and sputtered from the limiter can be pumped out of the plasma chamber, then the limiter can be utilized as a mechanism for removing impurities

High Density Divertor

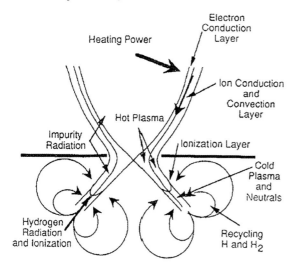

Figure 6.10 Processes taking place in the divertor.

from the plasma. The erosion of the limiter can be a major impurity problem, because of the proximity to the plasma, unless the edge of the plasma can be cooled (radiatively) sufficiently that the particle fluxes are incident on the limiter with energies below the sputtering threshold.

A more realistic depiction of a divertor configuration is given in Figure 6.9, and Figure 6.10 shows the many processes that go on in the divertor.

Problems

6.1. Estimate the flux of plasma ions to the wall from a tokamak plasma with ion density $2 \times 10^{20} \, \mathrm{m}^{-3}$, plasma radius $a = 1$ m, and particle confinement time $\tau_p = 1$ sec.

6.2. A D-T particle flux of $3 \times 10^{21} \, \mathrm{m}^{-2} \cdot \mathrm{s}^{-1}$ strikes the surface of a stainless-steel first wall. The average particle energy is 200 eV. Calculate the rate of impurity production in a tokamak with major radius $R = 5$ m and plasma radius $a = 1$ m. (Use Fe sputtering yields)

6.3. The density of stainless steel is $8 \, \mathrm{g/cm}^3$. Calculate the thickness and mass of wall eroded away in one full year of continuous operation for Problem 6.2. (Use $A = 55$ amu for stainless steel.)

6.4. Assume that a tokamak reactor plasma discharge terminates disruptively once every 1000 pulses and that the plasma internal energy goes to 10% of the total wall area in 10 msec. For a tokamak with $R = 5$ m, $a = 1$ m, $n = 2 \times 10^{20} \, \mathrm{m}^{-3}$, and $T = 20$ keV, what is the thickness of stainless-steel wall evaporated in 1 yr?

6.5. Calculate the bremsstrahlung, line, and recombination radiation power losses from a plasma with $n_e = 1 \times 10^{20}\,\mathrm{m}^{-3}$ and an iron impurity concentration of $n_z/n_i = 10^{-2}$ for an electron temperature of $T_e = 100\,\mathrm{eV}$.

6.6. Repeat Problem 6.5 for $T_e = 10\,\mathrm{keV}$.

6.7. The actual impurity concentration allowable with ignition is less than that shown in Figure 6.7 because of transport losses. Use the empirical scaling law $\tau_E(\mathrm{sec}) = 5 \times 10^{-21} n(\mathrm{m}^{-3}) a^2$ (m) to estimate from the power balance the maximum allowable concentration of Fe consistent with ignition in a D-T plasma at $T = 10\,\mathrm{keV}$ and $a = 1.5\,\mathrm{m}$.

6.8. Repeat Problem 6.7 for beryllium impurity.

6.9. Estimate the threshold energy for deuterons for physical sputtering of carbon and of tungsten.

7
Magnetics

7.1
Magnetic Fields and Forces

The basic laws that determine the magnetic fields and forces associated with a current-carrying element (Figure 7.1) are those of Biot–Savart

$$d\mathbf{B} = \mu_0 \frac{\mathbf{j} \times \mathbf{r}}{4\pi r^3} dV$$

which gives the differential magnetic field $d\mathbf{B}$ associated with a current density $\mathbf{j}$ in a differential volume, dV; of Lorentz,

$$d\mathbf{F} = \mathbf{j} \times \mathbf{B}\, dV$$

which describes the force on a differential volume element with current density $\mathbf{j}$ in a magnetic field $\mathbf{B}$; and of Ampere,

$$\oint d\mathbf{I} \cdot \mathbf{B} = \mu_0 I$$

which relates the line integral of the magnetic field around a closed loop to the enclosed current I.

7.1.1
Long Straight Conductors

In a long straight conductor,

$$\mathbf{j}dV \equiv \mathbf{j}dA\, d\mathbf{l} \equiv I d\mathbf{l}$$

where the incremental length $d\mathbf{l}$ is a vector in the direction of the current density. The field at a distance r from the conductor is obtained by integrating the Biot–Savart relation over the length of the conductor. For a long conductor, the result everywhere except near the ends is

Fusion: Second, Completely Revised and Enlarged edition. Weston M. Stacey
Copyright © 2010 WILEY-VCH Verlag GmbH & Co. KGaA, Weinheim
ISBN: 978-3-527-40967-9

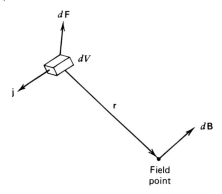

Figure 7.1 Current-carrying element.

$$\mathbf{B}(r) = \frac{\mu_0 I}{2\pi r} \hat{n}_\theta$$

where $\hat{n}_\theta$ is a unit vector in the poloidal direction.

Now consider two long straight conductors separated by a distance r. The differential force dF_2 on differential length dl_2 of conductor 2 carrying current I_2 due to the magnetic field $\mathbf{B}_1$ associated with the current I_1 in conductor 1 is found by using the above relation in the Lorentz force law,

$$d\mathbf{F}_2 = I_2 d\mathbf{l}_2 \times \mathbf{B}_1 = \frac{I_2 I_1}{2\pi r} \mu_0 d\mathbf{l}_2 \times \hat{n}_{\theta 1}$$

Similarly, the differential force on conductor 1 due to the current in conductor 2 is

$$dF_1 = I_1 d\mathbf{l}_1 \times \mathbf{B}_2 = \frac{I_2 I_1}{2\pi r} \mu_0 d\mathbf{l}_1 \times \hat{n}_{\theta 2}$$

From these results and Figure 7.2 it can be concluded that the force between two long straight conductors is attractive if the currents flow in the same direction in both conductors. It is straightforward to show that the force is repulsive if the currents flow

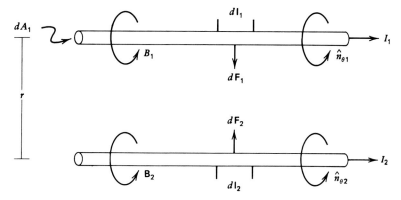

Figure 7.2 Long straight conductors.

in opposite directions in the two conductors. In either case the magnitude of the force is the same on both conductors.

7.1.2
Coaxial Loop Conductors

The magnetic field at any point ρ with respect to the center of a current-carrying loop (Figure 7.3) can be computed from the Biot–Savart law (or more readily, by constructing the vector potential). The radial component is

$$B_r(r,z) = \frac{\mu_0 Ikz}{4\pi(ar^3)^{1/2}}\left[-K(k) + \frac{(1-\frac{1}{2}k^2)}{(1-k^2)}E(k)\right]$$

and the axial component is

$$B_z(r,z) = \frac{\mu_0 Ik}{4\pi(ar)^{1/2}}\left[K(k) + \frac{(a+r)k^2-2r}{2r(1-k^2)}E(k)\right]$$

for the sense of current flow shown in Figure 7.3. On the axis of the loop ($r=0$), this reduces by L'Hôpital's rule to

$$B_r(0,z) = 0 \qquad B_z(0,z) = \frac{\mu_0 Ia^2}{2(a^2+z^2)^{3/2}}$$

Here K and E are elliptic integrals of the first and second kind, respectively,

$$E(k) = \int_0^{\pi/2} (1-k\sin^2\theta)^{1/2}d\theta$$

$$K(k) = \int_0^{\pi/2} (1-k\sin^2\theta)^{-1/2}d\theta$$

and

$$k \equiv \left[\frac{4ra}{z^2+(r+a)^2}\right]^{1/2}$$

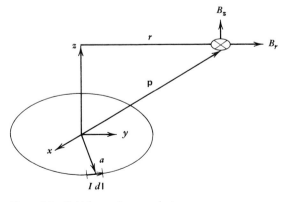

Figure 7.3 Field from a loop conductor.

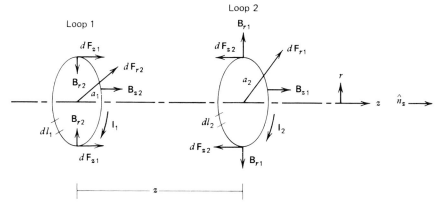

Figure 7.4 Coaxial current-carrying loops.

Now consider two coaxial loops of radius a_1 and a_2 with currents flowing in the same direction (into the paper at the bottom and out at the top in Figure 7.4). The radial magnetic field produced at loop 2 by the current in loop 1 is radially outward (z positive in the above expression for B_r), and the resulting differential axial force on loop 2

$$dF_{z2} = -B_{r1}(2)I_2 \oint dl_2 \hat{n}_z = -2\pi a_2 I_2 B_{r1}(2)\hat{n}_z$$

is directed toward loop 1 when the current is flowing in the same direction in the two loops. The radial magnetic field produced at loop 1 by the current in loop 2 is radially inward (z negative in the expression for B_r), and the resulting differential axial force on loop 1

$$dF_{z1} = B_{r2}(1)I_1 \oint dl_1 \hat{n}_z = 2\pi a_1 B_{r2}(1)I_1 \hat{n}_z$$

is directed toward loop 2 when the current is flowing in the same direction in the two loops. It is straightforward to show that the forces acting between the two loops would be repulsive when the currents flowed in the opposite direction. In either case, the magnitude of the force is the same on both loops.

The axial fields produced at loop 2 by loop I and at loop 1 by loop 2 are both in the positive direction for the current directions shown in Figure 7.4. The resulting differential radial forces are outward. If the current in either loop was reversed, so that the currents in the two loops flowed in opposite directions, the differential radial forces would be inward in both loops.

7.1.3
Ideal Torus

Consider an ideal thin-shell torus wound with N turns, each carrying current I, to form a continuous conducting sheath, as depicted in Figure 7.5. Using Ampere's law integrated around the loop at radius R relative to the major (z) axis of the torus leads to

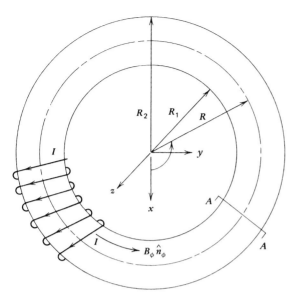

Figure 7.5   Ideal torus.

$$\mu_0 NI = 2\pi R B_\phi(R) \qquad R_1 < R < R_2, \quad z = 0$$

$$0 = 2\pi R B_\phi(R) \qquad R < R_1, \quad R > R_2, \quad z = 0$$

from which the axial, or toroidal, magnetic field within the torus can be written,

$$\mathbf{B}_\phi(r) = \frac{\mu_0 NI}{2\pi R}\,\hat{n}_\phi \qquad R_1 < R < R_2$$

Since B_ϕ vanishes immediately outside the torus, but has a finite value immediately inside the torus, its average value within the conducting sheath that forms the torus wall can be approximated as one-half the value immediately inside the torus. Then the differential force acting on a unit length of conductor (Figure 7.6) is given by the Lorentz law, using one-half the above expression for B_ϕ,

$$d\mathbf{F} = \frac{1}{2}\,Id\mathbf{l} \times \mathbf{B}_\phi = \frac{\mu_0 NI^2}{4\pi R}\,d\mathbf{l} \times \hat{n}_\phi$$

Denoting the value of R at the center of the torus by R_0, R can be written as

$$R = R_0\left(1 + \frac{r}{R}\cos\theta\right) \equiv R_0(1 + \varepsilon\cos\theta)$$

Using this expression in the force relation yields an expression for the differential force per unit length acting on each turn,

$$d\mathbf{F} = \frac{\mu_0 NI^2 d\mathbf{l} \times \hat{n}_\phi}{4\pi R_0(1 + \varepsilon\cos\theta)} = \frac{\mu_0 NI^2 d\ell}{4\pi R_0(1 + \varepsilon\cos\theta)}\,\hat{n}_r$$

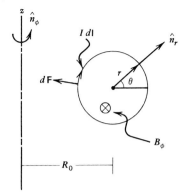

Figure 7.6 Force on ideal torus.

This differential force is everywhere outwardly normal to the surface of the torus (the $\hat{n}_r$ direction) and varies in magnitude as $(1 + \varepsilon \cos \theta)^{-1}$. This force can be decomposed into R and z components, as depicted in Figure 7.7.

Thus, there is a net inward (toward the major axis of the torus) force on each of the N turns, obtained by integrating around the turn,

$$\mathbf{F}_R = \oint d\mathbf{F}_R = \frac{\mu_0 N I^2}{4\pi R_0} \int_0^{2\pi} \frac{r \cos \theta \, d\theta}{1 + \varepsilon \cos \theta} \hat{n}_R = -\frac{\mu_0 N I^2}{2} \left[1 - \frac{1}{(1-\varepsilon^2)^{1/2}} \right] \hat{n}_R$$

This force must be reacted by some external support structure.

The net axial (z-directed) force integrated over a turn is zero, so that no external support is required. However, the axial force does produce an internal tensile force in the conductor. This internal tensile force may be estimated by considering the axial force over half of the turn to be reacted internally by the tensile forces $T_1 + T_2$, as depicted in Figure 7.8. Using the axial component of the differential force given above,

$$T_1 + T_2 = \frac{\mu_0 N I^2}{4\pi R_0} \int_0^{\pi} \frac{r \sin \theta \, d\theta}{1 + \varepsilon \cos \theta} = \frac{\mu_0 N I^2}{4\pi} \ln \left(\frac{1+\varepsilon}{1-\varepsilon} \right)$$

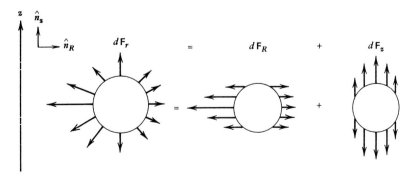

Figure 7.7 Decomposition of force on ideal torus.

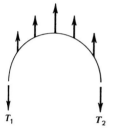

Figure 7.8 Tensile forces.

7.1.4
Solenoid

The results for the ideal torus can be applied to a continuously wound solenoid if the limit of a very large torus ($R \to \infty$) is taken and we imagine the torus to be cut and straightened into a cylinder of length $L = 2\pi R$. The previous expression for the axial field becomes (Figure 7.9)

$$\mathbf{B}_\phi \to \mathbf{B}_z = \mu_0 \left(\frac{NI}{L} \right) \hat{n}_z \qquad r < r_1$$

where N/L is now the number of turns per unit length, each carrying current I, and

$$\mathbf{B}_z = 0 \qquad r > r_2$$

If the current density is uniform across the thickness, $\Delta r = r_2 - r_1$, then the field varies as

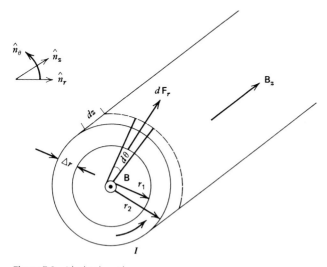

Figure 7.9 Ideal solenoid.

$$B_z(r) = \mu_0 \frac{NI}{L}\left(\frac{r_2 - r}{r_2 - r_1}\right) \qquad r_1 < r < r_2$$

within the conductor. Using this expression in the Lorentz force expression yields a radially outward force on the conductor,

$$d\mathbf{F}_r = \int j_\theta B_z dV\,\hat{n}_r = j_\theta \frac{\mu_0 NI}{L}\int_{r_1}^{r_2} dr\, r\, d\theta\, dz\left(\frac{r_2 - r}{r_2 - r_1}\right)\hat{n}_r$$

$$= \frac{1}{2}\mu_0\left(\frac{NI}{L}\right)^2 r_1\left(1 + \frac{\Delta r}{3r_1}\right)d\theta\, dz\,\hat{n}_r$$

where $j_\theta \Delta r = NI/L$ has been used. This differential force must be reacted by a tensile force $\mathbf{T}$ in the coil. The radial force balance on a differential element of the solenoid, depicted in Figure 7.10, is

$$2T\sin\frac{d\theta}{2} = dF_r$$

The tensile stress σ is related to the tensile force by

$$T = \sigma\, dA = \sigma\Delta r\, dz$$

from which follows an expression for the average tensile stress in a solenoid with N/L turns per unit length, each with current I,

$$\sigma = \mu_0\left(\frac{NI}{L}\right)^2\frac{1}{2}\left(\frac{r_1}{\Delta r} + \frac{1}{3}\right)$$

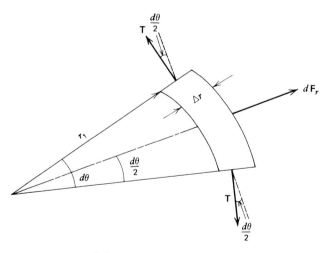

Figure 7.10 Force balance.

7.2
Conductors

7.2.1
Normal Conductors

Most of the magnets in fusion experiments to date have used copper as the conductor. Aluminum can also be used as a conductor. Copper and aluminum are referred to as normal conductors to distinguish them from superconductors, which are discussed in the next section.

The conductor is wound in one of the three generic configurations shown in Figure 7.11. It may consist of internally cooled hollow cables or tapes separated by electrical insulator, as shown in Figures 7.11a and b. The Bitter-coil configuration depicted in Figure 7.11c is constructed of radially or axially cooled disks, twisted to form a spiral, and separated by insulation. Normal conductors are usually cooled by water and operate at room temperature.

However, since their resistivity decreases substantially at reduced temperatures, the conductors are sometimes cooled by liquid nitrogen, in which case the terminology "cryogenic" magnet is used. The temperature dependence of the resistivity of copper is given near room temperature by

$$\rho_{Cu}(\Omega \cdot m) \simeq \{1.68 + 0.0068[T(K) - 293]\} \times 10^{-8}$$

The conductors in Figure 7.11 are sometimes placed in a steel jacket for added support.

The principal limitation on the use of normal magnets in fusion experiments and future reactors is the resistive power dissipation. For a magnet with N turns, each carrying current I, the resistive power loss in each turn is

$$P_T = I^2 R = I^2 \left(\frac{\rho L_T}{A_c} \right)$$

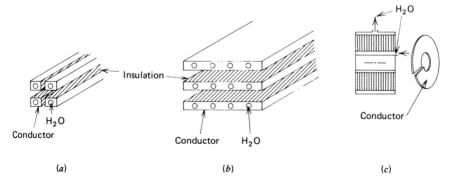

(a)   (b)   (c)

Figure 7.11 Normal conductor configurations: (a) single-hole hollow conductor; (b) multiple-hole hollow conductor; (c) Bitter conductor.

where ρ is the resistivity, L_T is the length of the turn, and A_c is the cross-sectional area of the conductor.

The magnetic field strength inside the conductor can be written

$$B = \mu_0 NIf_1$$

where f_1 is a geometric factor [$f_1 = \frac{1}{2}(2\pi R)^{-1}$ for an ideal torus, $f_1 \simeq \frac{1}{2}L^{-1}$ for a solenoid). Thus, the maximum field depends upon the maximum allowable power dissipation,

$$P = NP_T = \frac{1}{N}\left(\frac{B}{\mu_0 f_1}\right)^2 \frac{\rho L_T}{A_c}$$

The dissipated heat must be removed by water. Thus, if P_T is the dissipated heat in a turn of length L_T, and L_p is the perimeter of the coolant–conductor interface in a cross section of the magnet, then the heal removal condition is

$$\frac{P_T}{L_T} = L_p h(T_c - T_f) = q''L_p$$

where T_c and T_f are the conductor and fluid temperatures and h is the convective heat transfer coefficient between the conductor and the coolant. (Calculation of h from the properties of the coolant is discussed in Section 10.3). q'' is the heat flux per unit area from the conductor into the coolant.

The coolant (H_2O) mass flow rate required to remove the dissipated heat in each turn P_T is

$$\dot{W}(\text{kg/sec}) = \frac{P_T}{C_p \Delta T_f}$$

where C_p is the coolant specific heat and ΔT_f is the coolant temperature rise per turn. The pressure drop associated with this mass flow rate over a distance L_T in a flow area A_f is

$$\Delta p = \frac{f\rho_f}{2D}\left(\frac{\dot{W}}{\rho_f A_f}\right)^2 L_T$$

where D is the equivalent diameter of the flow area, ρ_f is the coolant density, and f is the Moody friction factor (which is evaluated in terms of the coolant properties in Section 10.3). The pumping power per turn is

$$P_p = \frac{\Delta p \dot{W}}{\rho \eta_p}$$

where η_p is the pump efficiency.

Combining these results, it is apparent that, for a given conductor cross section and length per turn and for a given coolant, there is a maximum value of the current for which the dissipative heat can be safely removed. The current capacity of a commercially available hollow copper conductor is given in Figure 7.12.

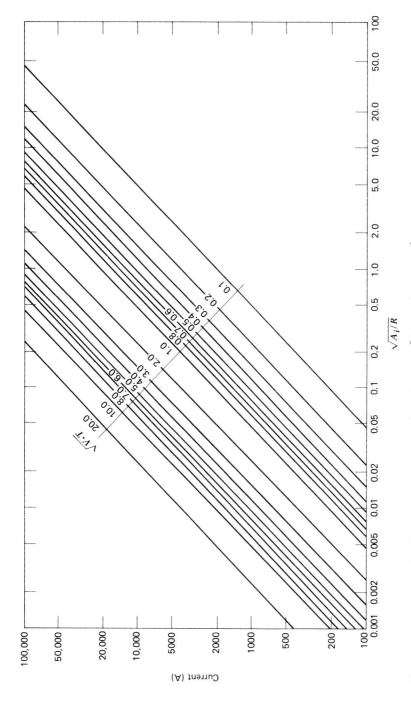

Figure 7.12 Current capacity of hollow copper conductor. A_i – hole area (in.^{2}); R – resistance (10^{-6} Ω/in.) at copper temperature; V – water velocity (ft/sec); T – temperature rise for water ($^\circ$F/in.). (Reprinted with permission of the Anaconda Co.)

At the field strengths (8–12 T) and dimensions projected for future fusion reactors, the resistive power dissipation would be hundreds of megawatts. This would make it very difficult to obtain a positive power balance on a reactor. Thus, increasing attention is being devoted to superconducting magnets.

7.2.2
Superconductors

Some materials have the rather amazing property that as their temperature is reduced to near absolute zero, their electrical resistivity will suddenly drop to essentially zero. Such a state is referred to as superconducting, and materials which exhibit this property are known as superconductors. The great incentive for the use of super-conductors in fusion magnets is the elimination of the huge dissipative power loss associated with normal conductors.

The superconducting state occurs within a relatively small region of temperature – magnetic field-current density phase space, outside of which the material returns to finite resistivity. This superconducting region of phase space is depicted in Figure 7.13 for three materials that are being considered for fusion magnets. NbTi is an alloy for which there exists a rather substantial basis of experience with large magnets. It is a ductile material, which facilitates its use as a magnet conductor. The only limitation of NbTi is its relatively small superconducting region, which places practical limits of ~8–10 T on the maximum magnetic field. This limitation of NbTi

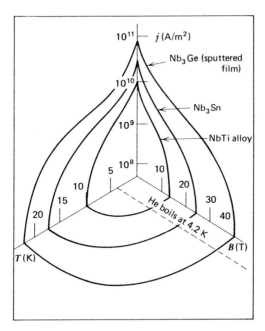

Figure 7.13 Superconducting regions in phase space. (Reproduced with permission from T. J. Dolan, *Fusion Research*, Vol. III. Pergamon Press, Elmsford, N.Y., 1982.)

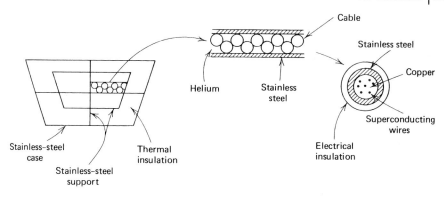

Figure 7.14 Superconducting conductor configuration.

has stimulated work on Nb_3Sn, which has a substantially larger superconducting region. Unfortunately Nb_3Sn is a brittle material, as a result of which there was until recently almost no large magnet experience with it. However, the recent development of a cable-type conductor configuration incorporating steel promises to ameliorate this concern, and Nb_3Sn has now been tested in large magnets. There is very little practical magnet experience with Nb_3Ge.

A representative configuration for a superconducting magnet conductor is depicted in Figure 7.14. Superconducting wires, composed of many superconducting filaments, are embedded in a copper (Cu) matrix, which in turn is wrapped in stainless steel (SS) and then wrapped with a spiral of tapelike electrical insulator to form the basic cable. The amount of superconductor is determined from the amount of current to be carried and from the upper limit on the current density (with some margin) indicated in Figure 7.13. The purpose of the copper is to provide stabilization of the superconducting state against unanticipated heating, and the purpose of the steel is to provide structural support to counteract the Lorentz ($I \times B$) forces that act on the conductor. Several cables may be combined (they are usually braided about each other) within a steel box to form a tape, with room left for liquid helium coolant. The helium either flows by natural convection or is pumped to obtain better heat removal; it is at 4.2 K at atmospheric pressure or colder if at reduced pressure. These tapes are built up within some structural system which is surrounded by thermal insulation and finally an outer case.

7.2.3
Stability and Quench Protection

The superconducting wires and the copper matrix in which they are embedded constitute parallel electrical paths with resistances $R_{SC} = \rho_{SC} L_T / A_{SC}$ and $R_{Cu} = \rho_{Cu} L_T / A_{Cu}$. Under normal conditions $\rho_{SC} \to 0$, $R_{SC} \ll R_{Cu}$, and the current flows entirely in the superconductor. (The residual resistivity of copper at 4.2 K is about $10^{-6}\ \Omega$ cm.)

There are various processes (e.g., reduction in internal energy, loss of coolant, nuclear heating, frictional heating, ac losses, resistive heating) which can lead to an undesired local temperature increase in the superconductor which causes a local excursion out of the superconducting region in phase space. This is referred to as "going normal." When this happens, R_{SC} increases dramatically from zero to a finite value. If the copper were not present, the current in the superconductor would be governed by

$$L_{SC}\dot{I}_{SC} + R_{SC}I_{SC} = 0$$

which has the solution

$$I_{SC}(t) = I_{SC}(0)e^{-(R_{SC}/L_{SC})t}$$

During the finite decay time there would be substantial resistive heating in the superconductor. This heat would conductively enlarge the nonsuperconducting, or normal, spatial region. Protection circuits are necessary to prevent this process from leading to melting and destruction of the magnet.

The presence of the copper is intended to ensure against the propagation of a normal region. When the superconductor goes normal locally, the resistance of the superconducting path becomes much greater than the resistance of the parallel copper path because of the much greater area of the latter ($\rho_{SC} \simeq \rho_{Cu}$, $A_{SC} \ll A_{Cu}$). If the area of the copper is sufficiently large and the provision for cooling is adequate then the local normal region can be cooled back down within the superconducting region in phase space and the superconducting state can be recovered. This process is referred to as cryogenic stabilization, and the necessary condition for its realization is

$$\alpha_s = \frac{\text{power dissipation}}{\text{heat removal}} = \frac{I^2 R}{q''lL_p} = \frac{I^2 \rho_{Cu}l/A_{Cu}}{q''lL_p} < 1$$

where l (which cancels out) is the length of the affected region, q'' is the heat flux into the helium (which is limited to $q'' < 0.4 \, \text{W/cm}^2$ in liquid helium by nucleate boiling), and L_p is the perimeter of the coolant–conductor interface. This criterion determines the required amount of copper (i.e., A_{Cu}).

If the conductor specific heat is sufficient to absorb the heat energy without a significant increase in local temperature, so that the normal region does not grow, the magnet is adiabatically stable. Small superconducting filaments, high copper thermal and electrical conductivity and high copper specific heat are required for adiabatic stability, while high current density is undesirable.

If the normal region can not be stabilized, then the magnet must be protected from the consequences of a large local hot region – overheating, overstress, overpressure and high voltages. To this end the voltages within the magnet are monitored to detect disturbances that could cause such a "quench" and to take action to make the magnet normal everywhere so as to distribute the heat and thereby limit the maximum hot-spot temperature and to couple the magnet with a secondary circuit with a dump resistor to resistively dissipate the stored energy outside the magnet.

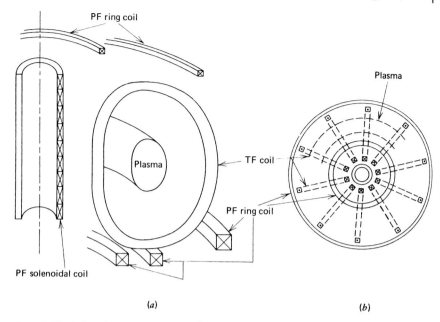

Figure 7.15 Tokamak magnet system configuration: (*a*) elevation view; (*b*) plan view.

7.3
Tokamak Magnet Systems

The magnet systems of a tokamak fall into two distinctly different categories, the poloidal field (PF) coil system and the toroidal field (TF) coil system. The PF coils are coaxial with the plasma, while the TF coils link, or wrap around, the plasma. A representative configuration is illustrated in Figure 7.15.

7.3.1
Poloidal Field Coils

The PF coils serve a number of functions in a tokamak. These functions determine the placement of the coils and the current carried.

The transformer action to induce the plasma current is provided primarily by a solenoidal ohmic heating (OH), or induction, coil. A changing current in this solenoidal OH coil provides a changing magnetic flux through the surface bound by the plasma ring, thus inducing an electromotive force V_p around the plasma ring, which induces a current

$$L_p \dot{I}_p + I_p R_p = V_p = -\dot{\Phi}$$

This process is depicted in Figure 7.16. The total flux change required to induce a current I_p^f in the plasma can be estimated by integrating this equation over the induction time. Neglecting the plasma resistance,

Transient Electromagnetics

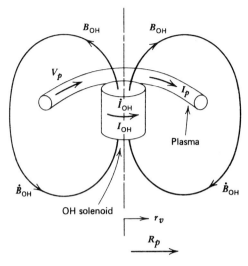

Figure 7.16 Current induction.

$$\Delta\Phi_{ind} = \int_0^{t_f} \dot{\Phi}dt' = L_p I_p^f \simeq \mu_0 R_p \left[\ln\left(\frac{8R_p}{a} - 1.75\right)\right] I_p^f$$

where the plasma self-inductance has been approximated by that of a loop conductor. The flux change required to overcome the resistive losses frequently turns out to be comparable to the inductive flux swing, so that a reasonable approximation is

$$\Delta\Phi = \Delta\Phi_{res} + \Delta\Phi_{ind} \simeq 2\Delta\Phi_{ind}$$

Since the magnetic flux must pass through the core of the solenoid,

$$\Delta\Phi = \pi r_v^2 \Delta B_{OH}$$

(Actually, not all of the OH flux linking the plasma passes through the OH solenoid, which is a significant effect.) In a normal conducting OH solenoid, I_{OH} will increase from zero to some final value I_{OH}^f and B_{OH} will correspondingly increase from zero to B_{OH}^f, where

$$B_{OH}^f = \mu_0 \left(\frac{N I_{OH}^f}{L}\right)$$

so that $\Delta B_{OH} \simeq B_{OH}^f$. In a superconducting OH solenoid, I_{OH} will increase from $-I_{OH}^f$ through zero to $+I_{OH}^f$, so that $\Delta B_{OH} \simeq 2B_{OH}^f$. Technological considerations probably will limit B_{OH}^f to about 13 T in superconductors and somewhat higher in normal conductors.

Ideally the magnetic flux produced by the OH solenoid links the plasma, but does not actually pass through the plasma. On the other hand, there is another set of

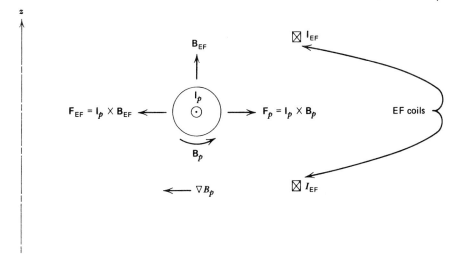

Figure 7.17 Force balance equilibrium produced by EF coil.

poloidal field coils, known as the equilibrium field (EF) coils, whose function involves producing a field that interacts with the plasma current to produce various desirable effects.

One principal function of the EF coils is to provide a radially inward force on the plasma to balance the net outward force produced by the interaction of the plasma current with its self-field B_p. This is depicted in Figure 7.17. The magnitude of the field B_{EF}, that must be produced at the plasma is

$$B_{EF} \simeq \frac{\mu_0 I_p}{4\pi R_p} \left[\ln\left(\frac{8R_p}{a}\right) - 1.5 + \beta_p + \frac{1}{2} l_p \right]$$

where l_p is the plasma internal inductance and β_p is the value of β in the poloidal field.

Another function of the EF coil system is to provide vertical elongation of the plasma, which leads to higher β limits against instabilities and better energy confinement. Elongation can be provided by a set of coils above and below the plasma which "pull" the plasma vertically. Imagining the upper and lower coils each interacting with half of the plasma current, the mechanism is illustrated in Figure 7.18a. If the current I_E in the elongation coils is sufficiently large, a null in the field will be formed and the field lines outside of the lines that pass through the null, known as the separatrix, are diverted out of the plasma chamber, that is, a poloidal divertor magnetic field configuration is formed, as shown in Figure 7.18b.

In addition to elongating the plasma, it is also advantageous to give it a D shape. This can be accomplished by "pushing" and/or "pulling" on the plasma, as depicted in Figure 7.19. The two coils on the left push harder on the inner (left) part of the plasma than the outer (right) part of the plasma, and the coil on the right pulls harder on the outer part than on the inner part of the plasma, in both cases because the field B_s produced by the coil is strongest near the coil.

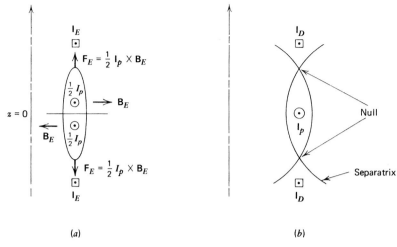

Figure 7.18 Elongation and separatrix formation: (a) elongation; (b) divertor.

A separately controllable set of coils can be employed for each of the functions of the PF coil system. However, the equilibrium, elongation, and shaping functions are similar in nature and must each be performed without interfering with the others. So it is common to think in terms of an EF coil system which performs these combined functions and an OH coil system which induces and maintains the plasma current.

7.3.2
Toroidal Field Coils

The principal function of the TF coil system is to provide a toroidal field to stabilize the plasma. This function is performed by a toroidal array of TF coils (see

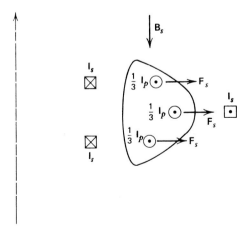

Figure 7.19 Plasma shaping.

Figure 7.15*b*). In present experiments there are a large number (~30) of coils that closely surround the plasma. In future reactors, shielding and access requirements will certainly lead to fewer (~10), larger coils. A cross-sectional view of a representative configuration is shown in Figure 7.20.

The forces that were discussed in Section 7.1.3 for an ideal torus are also predominant in a TF coil system composed of discrete coils. There is an inward, "centering" force that must be reacted by a support cylinder, and a tensile force must be borne by the conductor structure.

The toroidal field in the plasma is given, to a good approximation, by the expression derived in Section 7.1.3,

$$B_\phi(R) = \frac{\mu_0 NI}{2\pi R}$$

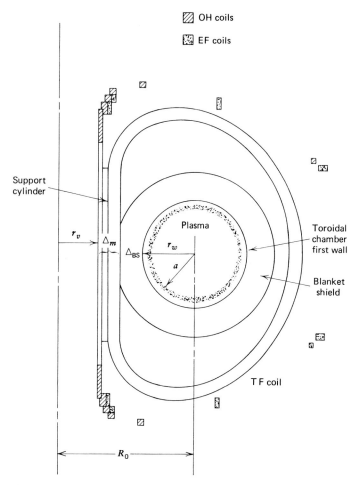

Figure 7.20 Tokamak nomenclature.

where now N is the number of discrete coils, each carrying current I. This expression may be used to relate the field at the center of the plasma $B_\phi^0 \equiv B_\phi(R_0)$ to the field at the coil $B_{TFC} \equiv B_\phi(R = r_v + \Delta_m)$,

$$B_\phi^0 = B_{TFC}\left(\frac{r_v + \Delta_m}{R_0}\right) = B_{TFC}\left(1 - \frac{r_w + \Delta_{BS}}{R_0}\right)$$

where Δ_{BS} is the blanket-shield thickness. Thus, there is a geometric reduction of the field at the center of the plasma with respect to the field at the coil. In future reactors, where shielding will be important, this reduction can approach a factor of 2.

The largest TF coils that have been built are those for JET, which are normal conducting, have an inner bore of 3.1×4.9 m, and will operate at $B_{TFC} = 6$ T. The largest superconducting TF coils is the ITER TF model coil (dimension 3×4 m). The yin-yang superconducting coil for MFTF-B, which has been successfully tested, has a typical dimension of 5 m and operates at 7.5 T. Future tokamak reactors are thought to require TF coils that are about 6.5×9.0 m and which operate at fields of 10–12 T. The ITER superconducting coils are discussed in a later section.

7.3.3
Coil Interactions

The interaction among the various coils in a tokamak magnet system is quite complex and of the utmost importance in determining the structural support requirements for the magnet system. Interactions among the discrete TF coils, between the TF coils and the PF coils, and among the discrete PF coils are all important.

Consider first the interaction among the discrete coils in the TF coil system, as depicted in Figure 7.21. Because of the discrete nature of the TF coils, radial (B_R) and axial (B_z) field components are produced, as well as the desired toroidal (B_ϕ)

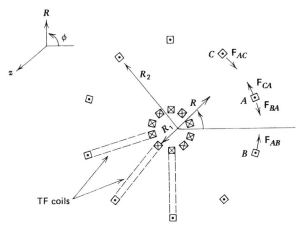

Figure 7.21 Interaction among discrete TF coils.

component. With respect to the figure, the B_R and B_z field components produced by coil B will interact with the current in coils C and A to produce attractive (toward coil B) forces $\mathbf{F}_{AB}$ and $\mathbf{F}_{AC}$ on these coils, and so on for the other coils in the array. (This is an extension of the result obtained previously of attractive forces between coaxial loops carrying currents in the same direction.) The magnitude of these forces is greatest at the inboard leg ($R = R_1$) where the coils are closest together, and weakest at the outboard leg ($R = R_2$) where the coils are furthest apart.

When all the coils carry the same current and are symmetrically located, the net force on each coil is zero (i.e., $|F_{BA}| = |F_{CA}|$ etc.). However, if the TF coil currents become unbalanced (e.g., due to failure of a coil) or the spacing becomes nonsymmetrical, then there would be a net toroidal force acting on a TF coil.

The discrete nature of the TF coils also results in radial and axial field components being produced at the location of a poloidal ring coil. These fields alternate in direction and magnitude along the poloidal coil, as depicted in Figure 7.22, resulting in alternating axial and radial forces acting along the ring coil,

$$d\mathbf{F}_z = \mathbf{I}_{PF} \times \mathbf{B}_{RT} dl$$

$$d\mathbf{F}_R = \mathbf{I}_{PF} \times \mathbf{B}_{zT} dl$$

The net force integrated around the PF coil is zero, but the coil must be supported to react the local force distribution.

The PF coils exert attractive or repulsive forces on each other, depending upon whether the currents are flowing in the same or opposite directions, respectively. These forces are of the type described in Section 7.1 for two coaxial ring conductors.

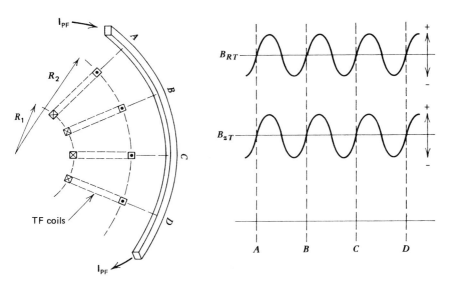

Figure 7.22 Fields produced by TF coils on PF coils.

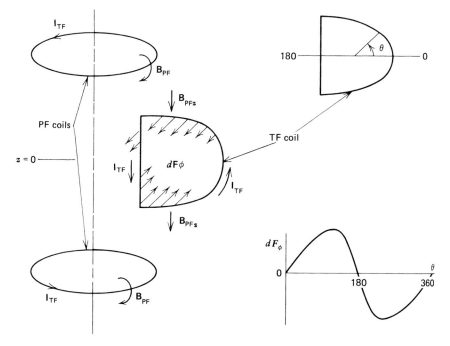

Figure 7.23 Overturning torque produced by PF coil system on TF coil.

The interaction of the axial component of the poloidal field with the current in a TF coil produces a torque about the midplane to act on the TF coil. At any instant during the cycle, the axial component of the poloidal field, B_{PFz}, is axially symmetric and thus has the same magnitude and direction at corresponding points above and below the $z=0$ plane. (This statement only applies to vertically symmetric plasmas.) The toroidal force

$$d\mathbf{F}_\phi = \mathbf{I}_{\mathrm{TF}} \times \mathbf{B}_{\mathrm{PFz}} dl$$

produced by the interaction of this axial field with the current in the TF coil is oppositely directed above and below the midplane (Figure 7.23), resulting in an overturning moment being exerted on the TF coil.

7.4
Structural Design Criteria

The structural support for the magnet systems in a fusion device must be designed to react the various static and dynamic loadings produced by gravitational, pressure thermal, and EM forces. The design criteria for the static loading must provide a safe margin against ductile rupture or gross yielding due to overloading. The ASME[1]

---

1) American Society of Mechanical Engineers.

criteria are

> primary tensile stress $< \alpha S_m$

and

> primary tensile plus bending stress $\leq \beta S_m$

where S_m is the lesser of two thirds yield stress or one-third ultimate stress at operating temperature. The parameters α and β are 1.0 and 1.5 under normal operation and 1.5 and 2.25 under abnormal operation.

The dynamic loadings due to pulsed magnetic fields lead to fatigue failure and to failure due to the growth of initially undetectably small cracks. The fatigue lifetime criterion is specified in terms of the allowable strain ε versus the number of cycles N_d of pulsed operation – ε is a decreasing function of N_d – for which safe operation can be assured. Thus, fatigue lifetime places a limit on the maximum allowable stress $\sigma^{max} = E\varepsilon^{max}(N_d)$ for a given number of cycles, where E is Young's modulus.

Another limit on maximum allowable stress is set by the growth of cracks under cyclic loading. This limit is dependent upon the stress ratio f of the cyclic stress to the total (cyclic plus static) stress amplitudes and upon the initial flaw size.

The yield stress and the ultimate stress for 316 stainless steel at 4 K are 125 and 200 ksi (875 and 1400 MPa). The allowable design stress, based upon the static loading criteria and the crack growth criterion, is shown in Figure 7.24 as a function of the stress ratio and the number of cycles, for a plausible smallest detectable flaw of 0.2 in. diameter. The crack growth criterion is quite limiting, much more so than the fatigue criterion, which is not shown.

7.5
ITER Magnet System

The ITER superconducting magnet system will use the cryogenically stabilized Nb_3Sn "cable in conduit" conductor shown in Figure 7.25 for both the toroidal field (TF) and central solenoid (CS) magnets.

The ITER toroidal field coil design is illustrated in Figure 7.26, and the major parameters are given in Table 7.1.

A major R&D program was carried out for the Nb_3Sn conductors from 1992–2002, culminating with the testing of two model coils, the Central Solenoid Model Coil (CSMC) and the Toroidal Field Model Coil (TFMC). Since the critical current density is sensitive to longitudinal strain, special manufacturing processes were developed to limit any strain introduced during manufacturing. Relevant lengths of the conductor were tested at relevant current and temperature in a 13 T magnetic field. The design of the ITER conductors was modified slightly based on the results of these tests.

A similar development program was carried out for the NbTi conductors to be used in the ITER equilibrium field (EF) coils. A tendency for premature quench at relatively low currents was discovered, which led to a design revision.

The principal parameters of the ITER magnet system are given in Table 7.2.

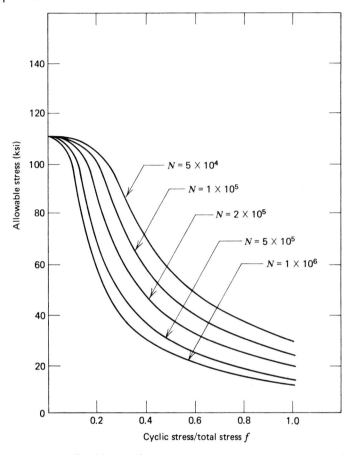

Figure 7.24 Allowable stress for 316 LN stainless steel at 4 K as determined by fracture mechanics. Initial flaw diameter $2a_0 = 0.2$ in; safety factor ≥ 2 on stress, ≥ 4 on cycles. (Reproduced with permission from W. M. Stacey, Jr *et al.*, "U.S. INTOR", Ga. Inst. Tech. report, 1980.)

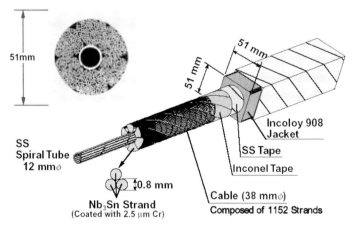

Figure 7.25 ITER "cable in conduit" conductor. (Reproduced with permission from ITER website.)

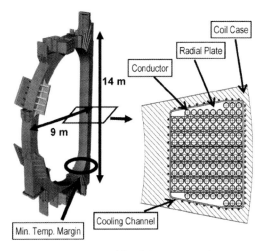

Figure 7.26 ITER toroidal coil dimension and cross-section (Reproduced with permission from ITER website.)

Table 7.1 Major parameters of conductors and TF Coils for analysis.

Operating Current	68 kA
Maximum Field	11.8 T
Operating Temperature at Peak Field	5.0 K
Discharge Time Constant	11 s
Strand Diameter	0.82 mm
Cu/non-Cu Ratio in Strand	1
Cabling Layout	$3 \times 3 \times 4 \times 4 \times 6$
Number of SC Strands	900
Number of Cu Strands segregated	288
Cable Diameter	40.5 mm
Central Tube OD × ID	9×7 mm
Local Void Fraction in Bundle	33.2%
Jacket External Dimension	43.7 mm

Table 7.2 Principal parameters of the ITER magnet system.

Number of TF Coils	18
Magnetic Energy in TF Coils (GJ)	41
Max. Field in TF Coils (T)	11.8
Centering Force per TF Coil (MN)	403
Vertical Force per Half TF Coil (MN)	205
TF Electrical Discharge Time Const. (s)	11
CS Peak Field (T)	13.0
Total Weight of Magnet System (tonne)	9000

Problems

7.1. Calculate the force per unit length acting between two long straight conductors, each carrying 1 MA of current and separated by 1 m.

7.2. Prove that the differential forces acting between two long straight conductors are repulsive when the currents in the two conductors are in opposite directions.

7.3. Prove that the differential axial forces acting between two coaxial loops are the same magnitude on both loops.

7.4. Calculate the axial and radial differential forces acting on two coaxial loop conductors, each carrying a current of 2 MA in the direction shown in Figure 7.4. The two coils have radii of $a_1 = a_2 = 5$ m and are separated axially by 2 m.

7.5. Calculate the net inward force acting on each turn of an ideal torus with 100 turns of minor radius $r = 4$ m, each carrying 1 MA of current. The major radius of the torus is $R_0 = 5$ m.

7.6. Calculate the tensile force acting within each turn in the previous problem.

7.7. Calculate the tensile stress in a long solenoidal magnet of inner radius 1.5 m and thickness 0.5 m that produces an axial field of 8 T.

7.8. Calculate the resistive power dissipation in the magnet of Problem 7.7 if the conductor is copper at room temperature and the solenoid is 5 m in height.

7.9. Calculate the upper limit on $A_{Cu}L_p$, the product of the copper area times the "wetted perimeter", that is required to ensure cryostability in a superconducting magnet carrying a current of 2 MA and operating at 4.2 K.

7.10. Consider two coaxial circular loop conductors with radii $R_1 = 2$ m and $R_2 = 5$ m separated axially by 1 m. Both conductors have 1 turn. Calculate the electromotive force that is induced in conductor 2 by a current change in conductor 1 of $\dot{I}_1 = 1 \times 10^6$ A/sec.

7.11. Estimate the total flux swing $\Delta\Phi$ required to induce and maintain a current of 7 MA in a tokamak plasma with $R = 5$ m and $a = 1.2$ m.

7.12. For Problem 7.11, estimate the minimum allowable value of the OH solenoid radius r_ν if the maximum allowable field is $B_{OH}^f = B_{OH}^{max} = 8$ T.

7.13. A tokamak with $r_\nu + \Delta_m = 2$ m is designed to produce $B_{TFC} = 10$ T in ten coils. Calculate the current in each coil.

7.14. If $\Delta_{BS} = 1$ m and $r_w = 1.4$ m in Problem 7.13, calculate the field at the center of the plasma.

8
Transient Electromagnetics

8.1
Inductance

In this section the basic concept of inductance is discussed as background for the subsequent discussion of transient EM coupling. Self-inductance and mutual inductance are considered.

First, the concept of self-inductance is developed by considering the stored energy in the magnetic field that is associated with a unit length of current-carrying conductor, as depicted in Figure 8.1. Letting the conductor lie along the z axis, the energy in the magnetic field associated with the incremental length l is

$$\frac{E}{l} = \int_{-\infty}^{\infty} dx \int_{-\infty}^{\infty} dy \frac{B^2}{2\mu_0}$$

For N turns each carrying current I, the magnetic field can be represented by

$$\mathbf{B} = \mu_0 NI[F_x \hat{n}_x + F_y \hat{n}_y]$$

where F_x and F_y are geometric functions depending on the conductor shape and the current density distribution. Thus,

$$\frac{E}{l} = \frac{1}{2}\mu_0(NI)^2 F_L$$

where F_L is a (dimensionless) geometric factor obtained by substituting the above form for $\mathbf{B}$ into the integral.

Now by definition, the self-inductance L for a linear system satisfies

$$E = \frac{1}{2}LI^2$$

Comparing these expressions leads to a prescription for the self-inductance,

$$\frac{L}{l} = \mu_0 N^2 F_L$$

Fusion: Second, Completely Revised and Enlarged edition. Weston M. Stacey
Copyright © 2010 WILEY-VCH Verlag GmbH & Co. KGaA, Weinheim
ISBN: 978-3-527-40967-9

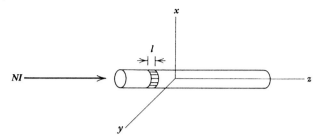

Figure 8.1 Current-carrying conductor.

For example, the self-inductance for a circular loop conductor of radius R and cross-sectional area A with N turns is

$$L = \mu_0 N^2 R \left[\ln \left(\frac{8R}{\sqrt{A/\pi}} \right) - 1.75 \right]$$

The mutual inductance between circuits b and c in Figure 8.2 can be thought of as the magnetic flux linked by circuit c per unit current flowing in circuit b when no current flows in circuit c. The magnetic field produced by the current in circuit b can be expressed in terms of the vector potential $\mathbf{A}_b$,

$$\mathbf{B}_b = \nabla \times \mathbf{A}_b$$

Here, as we have seen,

$$\mathbf{A}_b = \frac{\mu_0}{4\pi} \oint_b \frac{I_b d\mathbf{l}}{\rho}$$

where the integral is around circuit b and ρ is the distance from a point on the circuit b to the field point at which A_b is calculated.

The magnetic flux linking circuit c due to the current in circuit b is

$$\Phi_{bc} \equiv \int_{S_c} \mathbf{B}_b \cdot d\mathbf{s} = \int_{S_c} \nabla \times \mathbf{A}_b \cdot d\mathbf{s} = \oint_c \mathbf{A}_b \cdot d\mathbf{l}$$

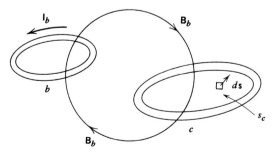

Figure 8.2 Magnetic flux linkage.

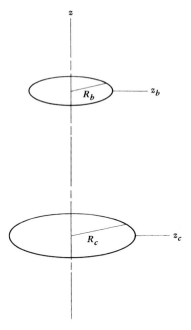

Figure 8.3 Coaxial conductors.

Now the mutual inductance of circuit c with respect to circuit b is defined as

$$M_{bc} \equiv \frac{\Phi_{bc}}{I_b} = \frac{1}{I_b} \oint_c \mathbf{A}_b \cdot d\mathbf{l}$$

The mutual inductance between two coaxial circular loop conductors with N_b and N_c turns is (Figure 8.3)

$$M_{bc} = \frac{2\mu_0 N_b N_c (R_b R_c)^{1/2}}{k} \left[\left(1 - \frac{1}{2} k^2\right) K(k) - E(k) \right]$$

where K and E are complete elliptic integrals of the first and second kinds, respectively, (Section 7.1.2), and

$$k \equiv \frac{2(R_b R_c)^{1/2}}{[(R_b + R_c)^2 + (z_b - z_c)^2]^{1/2}}$$

8.2
Elementary Electric Circuit Theory

The transfer of energy into and out of tokamak coil systems (and energy storage devices) is described by electric circuit theory. The basic laws of circuit theory are:

Kirchhoff's Voltage Law: The sum of the voltage drops around any loop is zero.

Kirchhoff's Current Law: The algebraic sum of the currents entering any junction point, or node, is zero.

The voltage drops across various types of circuit elements are

resistance R	IR
inductance L	$\dot{I}L$
capacitance C	$\dfrac{1}{C}\displaystyle\int_0^t I\,dt' = \dfrac{Q}{C}$

where I is the current flowing in the positive sense of the voltage drop and Q is the charge.

The various elements in a pulsed electric system in a fusion device can be represented in terms of their inductance, resistance, and/or capacitance. For example, a normal conducting magnet coil can be represented by an inductance and a resistance, while a superconducting coil can be represented only by an inductance. An idealized circuit for a power supply (voltage source V) coupled to a normal conducting coil is shown in Figure 8.4.

The circuit equation is formulated from Kirchhoff's voltage law,

$$V(t) = L\,\dot{I}(t) + RI(t) \qquad I(0) = 0$$

where the initial condition corresponds to the switch S_1 being closed at $t=0$ to energize the coil. The solution to this equation is

$$I(t) = \frac{1}{L}\int_0^t e^{-(R/L)(t-t')}V(t')dt'$$

Thus, the current buildup in the coil depends upon the characteristics of the power supply and the R/L value of the coil. For $V(t) = V_0 = $ constant, this solution reduces to

$$I(t) = \frac{V_0}{R}\left[1 - e^{-(R/L)t}\right]$$

The current approaches an asymptotic value $I(\infty) = V_0/R$ with a characteristic time $\tau \equiv L/R$.

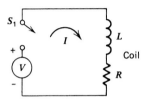

Figure 8.4 Idealized circuit.

The energy transferred into the coil as inductive energy in time t is

$$W_L = \int_0^t I(t')V_L(t')dt' = \int_0^t I(t')[L\,\dot{I}(t')]dt' = \frac{1}{2}LI^2(t)$$

and the energy dissipated resistively in the coil is

$$W_R = \int_0^t I(t')V_R(t')dt' = \int_0^t I(t')[I(t')R]dt' = R\int_0^t I^2(t')dt'$$

The voltage-generating characteristics of many energy storage systems are not constant, but rather depend upon the other elements in the circuit (the "load"). Two examples of this are discussed in the Section 8.5.

8.3
Coupled Electromagnetic Circuits

EM circuits in proximity to each other are coupled via magnetic flux linkage and mutual inductance. According to Faraday's law, a change in the magnetic flux through a surface bound by a circuit induces an electromotive force (voltage) $-\dot{\Phi}$ in that circuit,

$$-\frac{d\Phi}{dt} = -\int \frac{d\mathbf{B}}{dt}\cdot d\mathbf{s} = \int \nabla \times \mathbf{E}\cdot d\mathbf{s} = \oint \mathbf{E}\cdot d\mathbf{l} = V_{\text{ind}}$$

Thus, the equations governing the dynamics of N coupled circuits are

$$L_i\frac{dI_i}{dt} + R_iI_i + \frac{1}{C_i}\int_0^t I_i\,dt' = V_i - \sum_{j\neq i}^N \frac{d\Phi_{ji}}{dt}$$

$$= V_i - \sum_{j\neq i}^N \frac{d(M_{ji}I_j)}{dt} \quad i = 1,\ldots,N$$

Here V is any externally controllable voltage which is applied to the circuit from a power source, R is the resistance, L and M are the self- and mutual inductances, I is the current, and C is the capacitance of any energy storage device which may be in the circuit.

A change in the current in one circuit produces a change in the magnetic flux that links other coupled circuits, inducing a voltage which produces a change in the current in those other circuits. This phenomenon is used to induce a current in a tokamak plasma (which is a closed electric circuit) by changing the current in a set of induction coils to produce a changing magnetic flux that links the plasma ring. The vacuum vessel, first-wall, and structural systems, as well as the magnetic coils, constitute coupled EM circuits in a fusion device.

The response of any such system to an induced voltage is obtained by solving

$$L_i\,\dot{I}_i + R_iI_i = V_{\text{ind}} = -\sum_{j\neq i}\dot{\Phi}_{ji}$$

Assuming, for simplicity, that the induced voltage is constant, this equation has the solution

$$I_i(t) \simeq \frac{V_{\text{ind}}}{R}\left[1-e^{-(R_i/L_i)t}\right] + I_i(0)e^{-(R_i/L_i)t}$$

Thus, the characteristic time for the response of a circuit is

$$\tau_i = \frac{L_i}{R_i}$$

For two coupled circuits (without capacitive elements) it can be shown that the current in circuit 2 is

$$I_2(t) = \int_0^t [L_1 V_2(t-t') - M_{21} V_1(t-t')]\left[\frac{\omega_+ e^{-\omega_+ t'} - \omega_- e^{-\omega_- t'}}{\omega_+ - \omega_-}\right] dt'$$

$$+ \int_0^t R_1 V_2(t-t')\left[\frac{e^{-\omega_+ t'} - e^{-\omega_- t'}}{\omega_- - \omega_+}\right] dt'$$

where the characteristic frequencies are

$$\omega_\pm = \frac{(L_1 R_2 + L_2 R_1)[1 \mp \sqrt{1-4(L_1 L_2 - M_{21} M_{12})R_1 R_2/(L_1 R_2 + L_2 R_1)^2}]}{2(L_1 L_2 - M_{21} M_{12})}$$

The current in circuit 1 is obtained by interchanging the subscripts 1 and 2.

Induced currents in the vacuum vessel, structural elements, and the first wall usually are undesirable because of the associated resistive heating, the Lorentz forces $I \times B$ and the stray magnetic fields thereby produced. A sufficiently large induced current in the vacuum vessel or first wall effectively shields a tokamak plasma from the induction coils by creating a nullifying contribution to the electromotive force in the plasma. Thus, it is generally desirable that vacuum vessels, first walls, and structural elements be designed with a high resistance in any closed electric circuit.

8.4
Tokamak Transient Electromagnetics

Neglecting undesired current paths (vacuum vessel, first wall, etc.), the coupled EM circuits in a tokamak can be depicted as in Figure 8.5 and represented by the equations

$$L_{\text{OH}}\dot{I}_{\text{OH}} + R_{\text{OH}} I_{\text{OH}} + \frac{1}{C_{\text{OH}}}\int_0^t I_{\text{OH}} dt' + M_{\text{OH},p}\dot{I}_p = V_{\text{OH}}$$

$$L_{\text{EF}}\dot{I}_{\text{EF}} + R_{\text{EF}} I_{\text{EF}} + M_{\text{EF},p}\dot{I}_p = V_{\text{EF}}$$

$$L_p \dot{I}_p + R_p I_p + M_{\text{OH},p}\dot{I}_{\text{OH}} + M_{\text{EF},p}\dot{I}_{\text{EF}} = 0$$

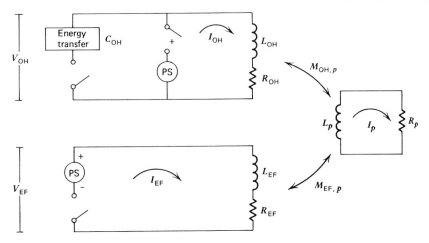

Figure 8.5 Circuit diagram for coupled EM circuits in a tokamak.

In writing these equations it has been assumed that the mutual inductance between the OH and EF coils is negligible, which will be the case in a good design. When superconducting coils are used, $R_{OH} = R_{EF} = 0$.

A typical operating sequence would be as follows: (i) At $t = 0$ gas is injected into the plasma chamber. (ii) In the OH circuit the switch is closed either to a capacitive-type energy transfer unit or to a power supply, in either case causing a changing OH current $\dot{I}_{OH}$. (iii) A plasma current is induced by mutual inductance with the OH system. (iv) The switch is closed on a controllable power supply in the EF circuit to produce $I_{EF}(I_p)$, which produces the equilibrium field that is required to position, elongate, and shape the plasma.

It is necessary to keep the OH power supply on, even after the final value of the plasma current is obtained, in order to offset resistive losses in the plasma and in the OH coil if the latter is not superconducting. As discussed in Chapter 1, the limitation on B_{OH} translates into a major limitation on achieving very long operating cycles, but noninductive methods to supplement or replace the action of the OH coils are being studied.

8.5
Energy Storage Systems

The energy storage requirements of fusion will be met primarily by inertial systems in the near term, with some use of capacitive storage systems. The future needs most probably will be met by inertial and inductive systems. The requirements upon the energy storage system are characterized by the total energy to be transferred, the energy transfer time, the voltage, the efficiency of the energy transfer, and the nature of the load (i.e., inductive, resistive). Some general features of energy storage systems for pulsed power supplies are given in Table 8.1.

Table 8.1 General features of pulse power supplies.

Type	Energy density (MJ/m³)	Transfer time (msec)	Cost ($/J)	Largest installation existing or expected
Capacitive	Low, 10^{-2}–10^{-1}	Very fast, 10^{-6} up	High, 0.10–0.25	10 MJ
Inductive	Medium, 10	Fast, 0.1–10	Medium, 0.01–0.10	100 MJ
Inertial (fast discharge homopolar)	High, 100	Fast, 1–100	Medium, 0.01–0.10	1000 MJ
Inertial (slow discharge homopolar)	High, 100	Medium, 1000–10 000	Low, 0.001–0.010	10 000 MJ
Inertial (MGF and SCR power supply)	High, 100	Medium, 1000–10 000	Medium, 0.01–0.10	10 000 MJ

Reproduced with permission from IEEE, H. H. Woodson *et al.*, *Proc. 6th Symp. Eng. Prob. Fusion Res.*, 1975.

8.5.1
Inertial Energy Storage and Transfer

Flywheels can be charged up with a motor, which draws power from the plant output or the utility line, to high rotation speeds. Flywheels rotating at high speeds can store inertial energy densities up to several hundred mega-joules per cubic meter, or ~0.1 MJ/kg. The rotating flywheel can be connected to an electric generator to produce a pulse of electric current, in the process decreasing the rotation speed, thus converting the inertial energy into electric energy. This type of inertial energy storage system is known as a motor–generator–flywheel (MGF) system and is widely used in present fusion experiments. The MGF set for the TFTR was capable of providing 600 MW for about 3 sec. A MGF set, consisting of four units, each storing 3.5 GJ available energy, was specified for the energy storage system of a typical tokamak reactor design.

The homopolar generator is an inertial energy storage device in which the flywheel is a simple rotor with no windings that can also serve as the armature for the generator. A disk-type homopolar generator is depicted in Figure 8.6. Energy from the utility line or plant output can be used to run a small motor which accelerates the rotating armature, or rotor. The energy is stored inertially until needed. The homopolar generator discharges by means of brushes on the shaft and on the rotor tapping the current that is driven by the radial electromotive force generated by the armature rotating azimuthally in an axial magnetic field.

The principle of the generator action of the inertial energy storage systems is based upon the fact that a medium moving with velocity **v** through space in which there is a magnetic field **B** will have induced in it an electric field

$$\mathbf{E} = \mathbf{v} \times \mathbf{B}$$

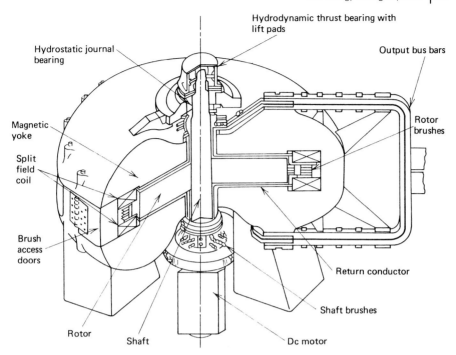

Hydrodynamic thrust bearing with lift pads

Hydrostatic journal bearing

Output bus bars

Magnetic yoke

Rotor brushes

Split field coil

Brush access doors

Return conductor

Shaft brushes

Rotor

Shaft

Dc motor

Figure 8.6 Homopolar generator energy storage system. (Reproduced with permission from T. J. Dolan, *Fusion Research*, Vol. III. Pergamon Press, Elmsford, N.Y., 1982.)

which for the homopolar generator of Figure 8.6 becomes

$$E_r = v_\phi B_z$$

The extended Faraday's law that applies to a moving medium is then

$$\nabla \times \mathbf{E} = -\frac{\partial \mathbf{B}}{\partial t} + \nabla \times (\mathbf{v} \times \mathbf{B})$$

A homopolar generator can be modeled as a capacitor with capacitance

$$C = J\left(\frac{2\pi}{\Phi}\right)^2$$

where J is the moment of inertia of the rotor and Φ is the total magnetic flux "cut" by the rotor.

The circuit equation for a homopolar generator coupled to a resistive and inductive load (e.g., a normal conducting coil) is

$$L\dot{I} + RI + \frac{1}{C}\int_0^t I \, dt' = 0$$

The solutions of this equation are well known and are characterized by the parameters

$$\lambda \equiv \frac{1}{2} R \sqrt{\frac{C}{L}} = \frac{\pi R}{\Phi} \sqrt{\frac{J}{L}}$$

and I_0, the maximum current in a circuit in which $\lambda = R = 0$, in Figure 8.7. The time axis in this figure is normalized to the time $\tau/4$ for a quarter cycle of an undamped $(R = 0)$ wave.

For a purely inductive load $(\lambda \simeq 0)$, such as a superconducting PF coil, the maximum current is

$$I_0 = \sqrt{\frac{2 W_0}{L}}$$

where W_0 is the energy stored initially in the homopolar generator and L is the self-inductance of the coil. The quarter-wave time is

$$\frac{\tau}{4} = \frac{\pi}{2} \sqrt{LC} = \frac{\pi^2}{\Phi} \sqrt{LJ}$$

These relations may be combined to obtain

$$\frac{\tau}{4} I_0 = \frac{\pi W_0}{V_0}$$

where $V_0 = \sqrt{2 W_0/C}$ is the initial voltage. This last relation shows that the product of the discharge time and the maximum current is limited by the initial energy and voltage; the maximum current and the discharge time can be traded off against each other.

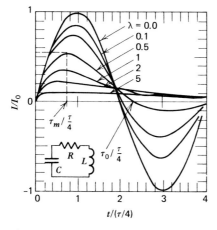

Figure 8.7 $I(t)$ in an *RLC* series circuit for various values of λ. (Reproduced with permission from IEEE, S. A. Mason and H. H. Woodson, Proc. 6th Symp. Eng. Prob. Fusion Res., 1975.)

For an underdamped $(0 < \lambda < 1)$ circuit with finite resistance, the maximum current I_m and time at which it occurs τ_m are given by

$$\frac{I_m}{I_0} = \exp\left[\frac{-\lambda}{\sqrt{1-\lambda^2}} \sin^{-1}\sqrt{1-\lambda^2}\right]$$

where I_0 is the maximum current given above for $\lambda = 0$,

$$\frac{\tau_m}{\tau/4} = \frac{2/\pi}{\sqrt{1-\lambda^2}} \sin^{-1}\sqrt{1-\lambda^2}$$

and the discharge time is

$$\frac{\tau_0}{\tau/4} = \frac{2}{\sqrt{1-\lambda^2}}$$

For a critically damped $(\lambda = 1)$ circuit with significant resistance, the maximum current is

$$\frac{I_m}{I_0} = e^{-1}$$

and the time of this maximum current is

$$\frac{\tau_m}{\tau/4} = \frac{2}{\pi}$$

For an overdamped $(1 < \lambda < \infty)$ circuit (essentially a resistive load), the maximum current is

$$\frac{I_{max}}{I_0} = (\lambda + \sqrt{\lambda^2 - 1})\exp\left(\frac{-\lambda}{\sqrt{\lambda^2 - 1}}\right)$$

and the time of this maximum current is

$$\frac{\tau_m}{\tau/4} = \frac{2/\pi\ln(\lambda + \sqrt{\lambda^2 - 1})}{\sqrt{\lambda^2 - 1}}$$

The fraction of the initial energy transferred to the inductive load is

$$\frac{W_L(t)}{W_0} = \frac{\frac{1}{2}L_L I_L^2(t)}{W_0} = \left[\frac{I_L(t)}{I_0}\right]^2$$

and the maximum energy transfer is

$$W_{max} = \left(\frac{I_m}{I_0}\right)^2 W_0$$

8.5.2
Capacitive Energy Storage and Transfer

Capacitor banks have been used in fusion experiments. They are ideally suited for very rapid discharge to provide pulsed power on the microsecond and faster time scale. However, the energy storage density is relatively small, and capacitors are not suited for providing the gigajoule level of pulsed energy that is required in present and future experiments. The discharge characteristics of capacitor banks were discussed in the previous section.

8.5.3
Inductive Energy Storage and Transfer

Energy can be stored inductively in electromagnets. A storage coil with inductance L_s carrying current I_0 stores energy $W_0 = \frac{1}{2} L_s I_0^2$. If the coil is superconducting, there is no resistive loss of the stored energy.

A simple circuit for transferring energy from a superconducting energy storage coil, with inductance L_s and initial current I_0, to a superconducting fusion device coil, with inductance L_L and zero initial current, is shown in Figure 8.8. The energy storage inductor is brought to full current by an external power source (not shown) with the switch S_1 closed, and then the external power source is switched off. The inductive energy $W_0 = \frac{1}{2} L_s I_0^2$ is stored in the storage coil until needed. There is no voltage across the storage coil during this storage period.

When the switch S_1 is opened, the commutation circuit (L_c, C) forces the current I_0 to transfer to the parallel path through the high-resistance shunt R. This produces an initial voltage $I_0 R$ across the three parallel paths (L_s, R, L_L) in Figure 8.8. The time-dependent currents are then given by (positive sense shown in figure)

$$I_s(t) = \frac{I_0(L_s + L_L e^{-t/\tau})}{L_s + L_L}$$

$$I_R(t) = -I_0 e^{-t/\tau}$$

$$I_L(t) = -\frac{I_0}{L_L + L_L}(L_s - L_s e^{-t/\tau})$$

and the time-dependent voltage is given by

$$V(t) = I_0 R e^{-t/\tau}$$

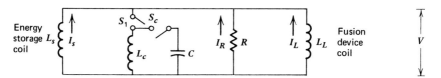

Figure 8.8 Single shunt resistor circuit for transferring energy between superconducting coils.

where the characteristic time constant is

$$\tau = \frac{L_s L_L}{R(L_s + L_L)}$$

The fraction of the energy initially stored in L_s which is transferred to the fusion device coil is

$$\frac{W_L}{W_0} = \frac{\frac{1}{2} L_L I_L^2}{\frac{1}{2} L_s I_0^2} = \frac{L_L L_s}{(L_s + L_L)^2}(1 - e^{t/\tau})^2 \underset{t \to \infty}{\longrightarrow} \frac{L_L L_s}{(L_s + L_L)^2}$$

Thus, the maximum energy transfer fraction is $W_L/W_0 = \frac{1}{4}$ when $L_s = L_L$.

The fraction of the energy initially stored in L_s which is dissipated in the resistive shunt is

$$\frac{W_R}{W_0} = \int_0^t \frac{I_R^2 R}{\frac{1}{2} L_s I_0^2} dt = \frac{L_L}{L_s + L_L}(1 - e^{-2t/\tau}) \underset{t \to \infty}{\longrightarrow} \frac{L_L}{L_s + L_L}$$

Thus, one-half the stored energy is ultimately dissipated in the shunt, independent of the value of the resistance, when $L_s = L_L$.

The fraction of the initial energy remaining in the energy storage coil is

$$\frac{W_s}{W_0} = \frac{\frac{1}{2} L_s I_s^2}{\frac{1}{2} L_s I_0^2} = \frac{(L_s + L_L e^{-t/\tau})^2}{(L_s + L_L)^2} \underset{t \to \infty}{\longrightarrow} \frac{L_s^2}{(L_s + L_L)^2}$$

Inductive energy storage with a resistive shunt transfer circuit is not very attractive because of its low energy transfer efficiency ($\leq 25\%$) and because the switch S_1 must be interrupted while carrying current I_0. Since the current is diverted to the resistive shunt path, a voltage is generated across S_1 equal to $I_0 R$. The efficiency of energy transfer between a superconducting storage coil and a superconducting fusion device coil can be improved dramatically by using a capacitive shunt transfer circuit. The simplest such circuit is illustrated in Figure 8.9.

The charging and storage procedure is the same as before, resulting in a current I_0 in L_s and no voltage for $t < 0$, with the switch S_1 closed and the other switches open. Transfer of energy is initiated at $t = 0$ by closing switch S_{c1} in the commutator loop, which connects C_1 across L_{c1}. Opening switch S_1 will then transfer the current in the parallel capacitive shunt path into C, charging up the capacitor (or homopolar generator) and generating a voltage across each coil,

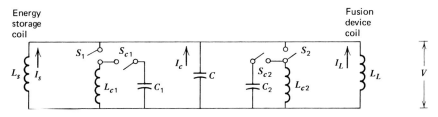

Figure 8.9 Single shunt capacitor circuit for transferring energy between superconducting coils.

$$V(t) = \frac{I_0}{\omega_0 C} \sin \omega_0 t$$

where

$$\omega_0 = \sqrt{\frac{L_L + L_s}{C L_s L_L}}$$

The currents are then given by

$$I_s(t) = \frac{I_0}{L_L + L_s}(L_s + L_L \cos \omega_0 t)$$

$$I_L(t) = -\frac{I_0 L_s}{L_L + L_s}(1 - \cos \omega_0 t)$$

$$I_c(t) = -I_0 \cos \omega_0 t$$

When the voltage reaches zero, the switch S_2 is closed. The current will continue to flow in L_L, which is now short-circuited through the commutative inductance L_{c2}.

The fraction of the initial energy transferred to the fusion device coil is

$$\frac{W_L(t)}{W_0} = \frac{\frac{1}{2} L_L I_L^2}{\frac{1}{2} L_s I_0^2} = \frac{L_s L_L (1 - \cos \omega_0 t)^2}{(L_L + L_s)^2}$$

If the transfer action is terminated (as discussed above) at

$$t_f = \pi \omega_0^{-1} = \pi \sqrt{\frac{C L_s L_L}{L_L + L_s}}$$

the energy transferred is

$$\frac{W_L(t_f)}{W_0} = \frac{4 L_L L_s}{(L_L + L_s)^2}$$

When $L_s = L_L$, 100% of the energy can be transferred.

Achieving this high-efficiency energy transfer is not without a price, however. The capacitive-type energy storage system used for the shunt must be able to store energy

$$W_c(t) = \frac{1}{2} C V^2(t) = \frac{L_L}{L_L + L_s} W_0 \sin^2 \omega_0 t$$

When $L_L = L_s$, the maximum energy stored in the "capacitor" (at $t = \pi/2\omega_0^{-1}$) is $W_c^{max} = \frac{1}{2} W_0$.

Energy is transferred from the fusion device coil back into the energy storage coil at the end of the cycle by an analogous procedure. Note that the circuit in Figure 8.9 is symmetric in the two coils and their associated commutative circuits.

Problems

8.1. Calculate the voltage needed to generate an asymptotic current of 2 MA in a room-temperature copper tokamak PF coil with radius 5 m and cross section $1\,m^2$.

8.2. Plot the current in the coil of Problem 8.1 as a function of time after a step increase of the voltage from zero to the required value.

8.3. A homopolar generator is used to energize a room-temperature copper tokamak PF coil with radius 4 m and cross section $0.8\,m^2$. Calculate the equivalent capacitance of a homopolar generator that would make the circuit critically damped.

8.4. The rotor of the homopolar generator in Problem 8.3 turns azimuthally in an axial magnetic field of 6 T and is of the disk type with a radius of 2 m. Calculate the rotor moment of inertia that is required to achieve the equivalent capacitance of Problem 8.3.

8.5. What initial energy must be stored in the homopolar generator of Problems 8.3 and 8.4 in order to produce a maximum current in the coil of 1.5 MA? At what time does this maximum current occur?

8.6. What is the maximum energy transferred to the coil in Problems 8.3–8.5?

8.7. A superconducting energy storage coil, with radius 7 m and area $1\,m^2$, is coupled to a superconducting PF coil, with radius 6 m and area $1.2\,m^2$, by a single shunt resistor circuit. Both coils have 10 turns each. Calculate the value of the shunt resistance that is required to achieve a characteristic time constant for energy transfer of 1 sec.

8.8. Plot the voltage and the currents in the energy storage and PF coils of Problem 8.7 as a function of time after the energy transfer is initiated, if the initial stored energy is 2 GJ.

8.9. Calculate the energy transferred to the PF coil and the energy dissipated in the resistive shunt after 10 sec in Problems 8.7 and 8.8.

8.10. Calculate the effective capacitance of a homopolar generator that could be used in a single shunt capacitor circuit to couple the two super conducting coils of Problem 8.7 and to achieve energy transfer in $t_f = \pi\omega_0^{-1} = 5$s.

8.11. Calculate the energy transferred to the PF coil in Problem 8.10 and the maximum energy transferred into the homopolar generator shunt if the initial energy in the inductive energy storage coil was 2 GJ.

8.12. Approximate the OH coil and the plasma in a tokamak as two coaxial loops with 200 and 1 turns, respectively, and radii of 1.5 m and 5.0 m, respectively. Calculate the self- and mutual inductances, if the areas of the plasma and the coil are $3\,m^2$ and $0.25\,m^2$.

8.13. Assume that the OH coil in Problem is superconducting and that the plasma has a cross-section area of $3\,m^2$ and a constant temperature of 500 eV. Calculate the plasma current induced by a voltage pulse in the OH coil of 100 V for 5 secs.

9
Interaction of Radiation with Matter

The 14-MeV fusion neutron from the D-T reaction is both a blessing and a curse. It carries with it 80% of the energy released by the fusion event when it leaves the plasma, thus providing a mechanism for distributing that energy over a sufficiently large volume that the heat removal problems are manageable. In addition, that neutron can be captured in lithium to produce tritium and thereby close the D-T fuel cycle. On the negative side, the high-energy neutron damages and activates the material surrounding the plasma and imposes a requirement for shielding to protect sensitive external components (e.g., superconducting magnets) and personnel.

9.1
Radiation Transport

The 14-MeV neutrons incident from the plasma upon the surrounding materials undergo a variety of reactions which alter their energy E, direction $\hat{\Omega}$, and number as they pass through the material. The principal reactions are summarized in Table 9.1.

Define $\psi(\mathbf{r}, E, \hat{\Omega})d^3r\,dE\,d\hat{\Omega}$ as the number of neutrons within incremental volume d^3r about point $\mathbf{r}$, with energy within incremental energy dE about energy E, and with direction within incremental solid angle $d\hat{\Omega}$ about direction $\hat{\Omega}$. (Henceforth we will speak of neutrons with energy E, and so on, to gain succinctness at the expense of precision.) The equation governing the distribution function ψ is then

$$\hat{\Omega} \cdot \nabla \psi(\mathbf{r}, E, \hat{\Omega}) + \sum_t(\mathbf{r}, E)\psi(\mathbf{r}, E, \hat{\Omega})$$
$$= S(\mathbf{r}, E, \hat{\Omega}) + \int dE' \int d\hat{\Omega}' \sum_s[\mathbf{r}, (E', \hat{\Omega}' \rightarrow E, \hat{\Omega})]\psi(\mathbf{r}, E', \hat{\Omega}')$$

The first term represents the net flow of neutrons with energy E and direction $\hat{\Omega}$ out of the incremental volume. The second term represents all nuclear reactions which can change a neutron's energy or direction (elastic and inelastic scattering reactions) or which can absorb the neutron into the nucleus [(n, γ), (n, α), (n, p), $(n, 2n)$, and so on]. The third term is the external source of neutrons, which for 14-MeV neutrons is the plasma fusion distribution. The last term is the source of neutrons within $dE\,d\hat{\Omega}$ due to reactions, primarily elastic and inelastic scattering but also including $(n, 2n)$

Fusion: Second, Completely Revised and Enlarged edition. Weston M. Stacey
Copyright © 2010 Wiley-VCH Verlag GmbH & Co. KGaA, Weinheim
ISBN: 978-3-527-40967-9

Table 9.1 Principal neutron reactions with materials.

Reaction	Symbol	Primary result
Elastic scattering	(n, n)	Neutron transfers kinetic energy to lattice atom
Inelastic scattering	(n, n')	
Capture	(n, γ)	Gamma emitted, $A \to A + 1$[a]
Capture	(n, p)	Proton emitted, $Z \to Z - 1$[a]
Capture	(n, α)	Alpha emitted, $Z \to Z - 2$, $A \to A - 3$[a]
Capture	$(n, 2n)$	Two neutrons emitted, $A \to A - 1$[a]
Capture	$(n, 3n)$	Three neutrons emitted, $A \to A - 2$[a]

a) Nuclear transmutation; A – atomic mass; Z – atomic number.

and $(n, 3n)$ reactions, of neutrons with other energies and directions which result in a neutron with energy E and direction $\hat{\Omega}$.

The macroscopic cross sections Σ_x are constructed from

$$\sum_x (\mathbf{r}, E, \hat{\Omega}) = \sum_i N_i(\mathbf{r}) \sigma_{xi}(E, \hat{\Omega})$$

where N_i is the number density of material i and σ_{xi} is the microscopic cross section for process x in material i. The microscopic cross sections are measured data and are tabulated in data files such as the evaluated nuclear data file (ENDF). The microscopic (n, α), (n, p), and $(n, 2n)$ cross sections for some candidate structural material nuclides are given in Figures 9.1 to 9.3 (1 barn $= 10^{-24}$ cm^2).

Gamma rays are produced by the neutron (n, γ) reactions. These gamma rays then undergo a number of reactions as they pass through the material. In the photoelectric reaction the gamma ray is destroyed and an energetic electron is emitted from the material. In the pair production process 1.02 MeV of the gamma ray energy is converted into the production of a positron – electron pair. The gamma ray can also undergo a change in energy and direction due to Compton scattering with a resultant absorption of energy. The transport of gamma rays can be represented by the same type of equation as written above for neutrons. Now Σ_t involves pair production, photo-electric reaction, Compton scattering and absorption; Σ_s involves pair production and Compton scattering; and the source term involves the neutron (n, γ) reaction rate.

Numerical techniques for solving the neutron and gamma transport equations are highly developed.

Because most of the nuclear reactions are independent of the neutron or gamma direction [i.e., the microscopic cross sections only depend upon energy $\sigma_{xi} = \sigma_{xi}(E)$], it is convenient to introduce the scalar flux

$$\phi(\mathbf{r}, E) \equiv \int \psi(\mathbf{r}, E, \hat{\Omega}) d\hat{\Omega}$$

Then the reaction rate for process x can be written

$$R_x(\mathbf{r}) = \sum_i N_i(\mathbf{r}) \int \sigma_{xi}(E) \phi(\mathbf{r}, E) dE$$

$$\equiv \sum_i N_i(r) \bar{\sigma}_{xi} \bar{\phi}(\mathbf{r})$$

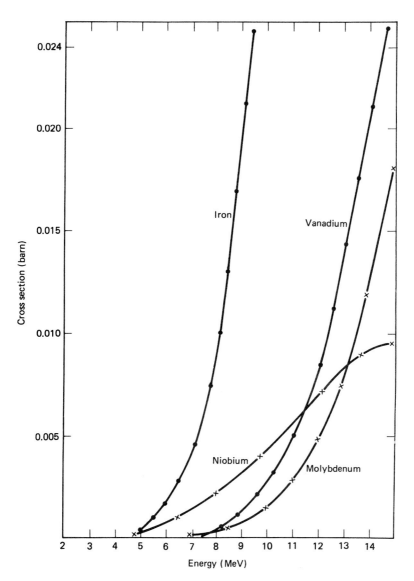

Figure 9.1 The (n, α) microscopic cross section for selected structural material nuclides. (Reproduced with permission from M.A. Abdou, Ph. D. thesis, University of Wisconsin, 1973.)

where we have defined the spectrum averaged microscopic cross section $\bar{\sigma}_{xi}$ and the integrated neutron flux

$$\bar{\phi}(\mathbf{r}) \equiv \int \phi(\mathbf{r}, E) dE$$

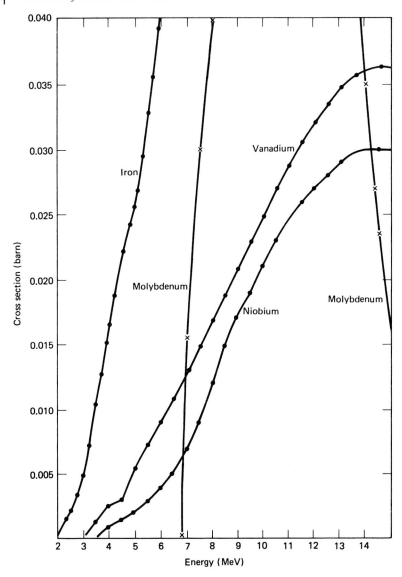

Figure 9.2 The (n, p) microscopic cross section for selected structural material nuclides. (Reproduced with permission from M.A. Abdou, Ph. D. thesis, University of Wisconsin, 1973.)

The magnitude of the neutron flux is attenuated rapidly as the neutrons pass through the material. The attenuation is roughly exponential with an "e-folding" distance on the order of 10 cm, depending, of course, upon the materials present. The energy degradation by scattering produces an energy distribution within the surrounding material that is significantly different from the monoenergetic 14-MeV fusion neutrons. Representative neutron spectra are shown in Figure 9.4.

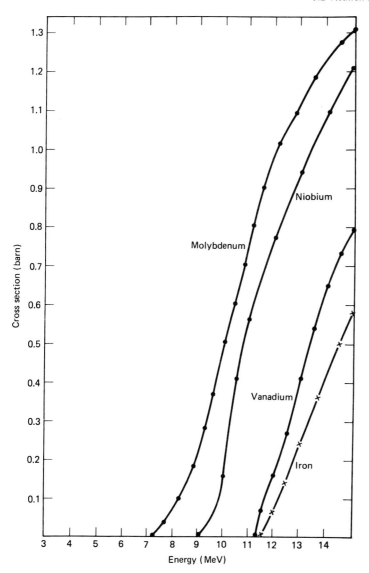

Figure 9.3 The (n, 2n) microscopic cross section for selected structural material nuclides. (Reproduced with permission from M.A. Abdou, Ph. D. thesis, University of Wisconsin, 1973.)

9.2
Neutron Multiplication

Most materials will undergo (n, 2n), (n, 3n), ... reactions with high-energy neutrons in which 2, 3, ... neutrons are emitted following neutron capture. There are a few materials of potential interest for fusion reactors in which the microscopic (n, 2n) cross section is significant, as indicated in Table 9.2.

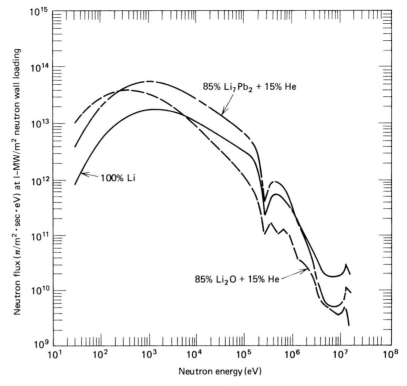

Figure 9.4 Comparison of neutron spectra in Li, Li_2O, and Li_7Pb_2 breeder systems at 0.3 m from first wall. (Reproduced with permission from D.L. Smith, Argonne National Lab report ANL/FPP–79–1, 1979.)

Table 9.2 Some neutron multiplying materials.

Material	$\sigma(n, 2n)$ at 14 MeV (barn[a])	$(n, 2n)$ Threshold (MeV)
Be	0.5	1.9
Pb	2.2	6.8
Bi	2.2	7.4
Zr	0.6	7.3

a) 1 barn $\equiv 10^{-24}\,cm^2$.

9.3
Nuclear Heating

The interaction of the neutrons and gamma rays with the lattice atoms of the material through which they pass converts the neutron and gamma energy to heat. The heat deposition at any spatial point due to neutrons or gammas can be expressed as

$$H(\mathbf{r}) = \int \phi(\mathbf{r}, E) \sum_j \sum_x N_j(\mathbf{r}) \sigma_{xj}(E) E_{xj}(E)$$

where ϕ is the neutron or gamma scalar flux, N_j is the number density of material j, and σ_{xj} is the microscopic cross section for reaction type x in material j.

For neutron heating the quantity $E_{xj}(E)$ is the total energy released in material j per reaction of type x induced by a neutron of energy E. This includes the kinetic energies of the recoil nuclei, charged particles emitted instantaneously, and charged particles emitted upon radioactive decay and other processes such as internal conversion. All of this energy is deposited locally.

The term

$$k_j(E) = \sum_x k_{xj}(E) = \sum_x \sigma_{xj}(E) E_{xj}(E)$$

is known as the kerma (kinetic energy released in materials) factor for material j and is tabulated.

The total energy deposited in the surrounding material by a fusion neutron is substantially greater than its original 14 MeV of kinetic energy because of exoergic nuclear reactions. The energy released (Q value) of several reactions in materials that might be found in a fusion reactor blanket is given in Table 9.3. In a typical blanket the energy deposited per 14-MeV neutron varies from ~17 to 24 MeV, depending upon the composition. Materials with large $(n, 2n)$ cross sections, which might be used for neutron multiplication, also tend to have exoergic reactions.

For gammas the kerma factors are defined similarly,

$$k_j^\gamma(E) = \sigma_{\text{pe},j} E + \sigma_{\text{pp},j}(E - 1.02) + \sigma_{\text{ca},j} E$$

where pe, pp, and ca refer to photoelectric, pair production, and Comptom absorption. In pair production 1.02 MeV (two electron masses) of the photon energy is unavailable for heating.

The spatial neutron and gamma heat deposition distributions calculated for a representative annular fusion reactor blanket are shown in Figure 9.5. The heating rates are normalized to an incident 14-MeV neutron power flux from the plasma of 1.25 MW/m², which is equivalent to a neutron particle flux of $5.6 \times 10^{17}\, n/m^2$ sec.

Table 9.3 Some exoergic reactions.

Material	Reaction	Q (MeV)
^{6}Li	(n, γ)	$+7.25$
^{6}Li	(n, α)	$+4.79$
^{7}Li	(n, γ)	$+11.34$
^{56}Fe	(n, γ)	$+7.80$
^{51}V	$(n, 2n)$	$+11.04$
^{51}V	(n, γ)	$+8.37$

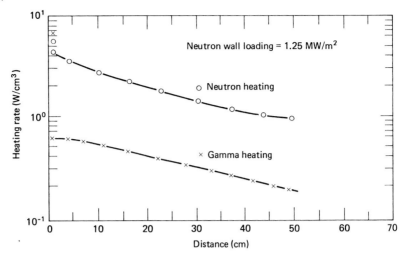

Figure 9.5 Neutron and gamma heating spatial distribution in a typical fusion reactor blanket. Zero corresponds to inner radius of first wall. (Reproduced with permission from B. Badger *et al.*, Univ. Wisconsin report UWFDM-68, 1974.)

9.4
Neutron Radiation Damage

9.4.1
Displacement and Transmutation

There are two primary responses of material subjected to bombardment by fast neutrons. Atoms are displaced from their lattice sites as a result of the energy imparted directly or indirectly to them by reactions with neutrons, producing displaced atoms, referred to as interstitials, and lattice vacancies, in pairs. Nuclear transmutations caused by (n, α) and (n, p) reactions produce helium and hydrogen gas in the material.

Neutron elastic, inelastic, $(n, 2n)$, and (n, γ) reactions can displace an atom from its lattice site. The displaced atom, known as a primary knock-on atom (PKA), in turn loses its energy by electronic excitation or nuclear energy transfers to surrounding atoms, causing further displacements in the process. The cumulative probability that a neutron with energy E displaces a lattice atom may be written

$$F(E) = F_{el}(E) + F_{in}(E) + F_{(n,2n)}(E) + F_{(n,\gamma)}(E)$$

where the subscripts refer to the reaction types, and the general form for $F_x(E)$ is

$$F_x(E) = \sigma_x(E) \int_{E_d}^{T_{max}} p_x(E, T)\nu(T)dT$$

Here σ_x is the microscopic cross section for reaction type x, $p_x(E, T)$ is the probability that an energy T is transferred to the PKA by a reaction of type x with a neutron of energy E, $\nu(T)$ is the number of subsequent displacements produced by a PKA with energy T, and E_d is the minimum energy required to displace an atom. All of these quantities depend upon the material.

The total displacement rate for material j is then

$$R_{dj}(\mathbf{r}) = \int F_j(E)\phi(\mathbf{r}, E)dE \equiv \sum_{dj} \bar{\phi}(\mathbf{r}) = N_j \bar{\sigma}_{dj} \bar{\phi}(\mathbf{r})$$

where ϕ is the neutron scalar flux, $\bar{\phi}$ is the total neutron flux integrated over energy, and $\bar{\sigma}_{dj}$ is the spectrum averaged displacement cross section for material j. The displacement is conventionally expressed as displacements per atom (dpa),

$$(\text{dpa})_j = \frac{R_{dj}}{N_j}\Delta t = \bar{\sigma}_{dj}\bar{\phi}\Delta t$$

where Δt is the time that the material is exposed to the neutron flux. Note that $\bar{\phi} \cdot \Delta t$ is the neutron fluence (n/m^2). Displacement distributions calculated for representative fusion reactor blankets are shown in Figure 9.6.

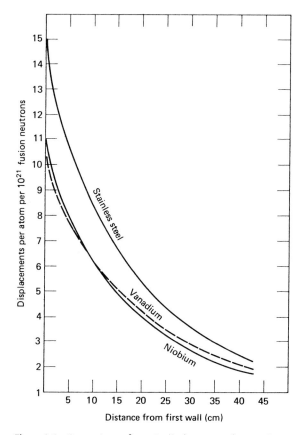

Figure 9.6 Comparison of atomic displacements for vanadium, niobium, and stainless steel. (Reproduced with permission from B. Badger *et al.*, Univ. Wisconsin report UWFDM-68, 1974.)

The helium or hydrogen production rate in material j due to nuclear transmutation can be calculated from

$$R_\alpha(\mathbf{r}) = \int N_j(\mathbf{r})\sigma_{(n,\alpha)j}(E)\phi(\mathbf{r}, E)dE$$

$$R_H(\mathbf{r}) = \int N_j(\mathbf{r})\sigma_{(n,p)j}(E)\phi(\mathbf{r}, E)dE$$

The helium and hydrogen production rate distributions per 14-MeV neutron leaving the plasma that were calculated for representative materials are shown in Figures 9.7 and 9.8.

The helium and hydrogen concentrations are commonly expressed in terms of the number of helium or hydrogen atoms per million lattice atoms, or in atomic parts per million (appm).

Table 9.4 summarizes the displacements and the helium and hydrogen concentrations that would result per 1 m^2 of surface area from a neutron fluence of 1 MW·yr/m^2 (4.48×10^{17} 14-MeV n/m^2) incident upon various materials.

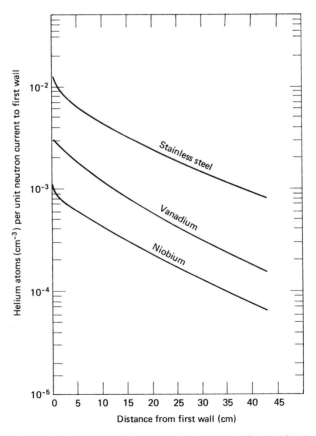

Figure 9.7 Comparison of helium production in vanadium, niobium, and stainless steel. (Reproduced with permission from B. Badger *et al.*, Univ. Wisconsin report UWFDM-68, 1974.)

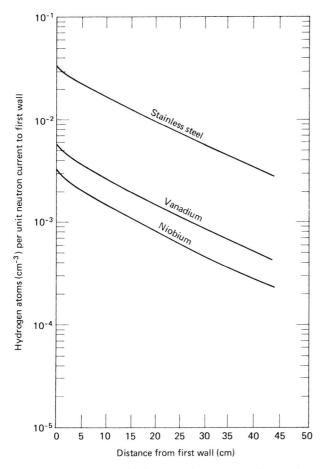

Figure 9.8 Comparison of hydrogen production in vanadium, niobium, stainless steel. (Reproduced with permission from B. Badger *et al.*, Univ. Wisconsin report UWFDM-68, 1974.)

Table 9.4 Primary response characteristics[a].

Material	Displacements per atom (dpa)	Helium (appm)	Hydrogen (appm)
316 stainless steel	10	200	540
Nb	7	24	79
Mo	8	47	95
V	12	57	100
C	10	2700	Small
Be		2800	130

a) Neutron fluence of 1 MW y/m^2.

Table 9.5 Physical and mechanical property changes.

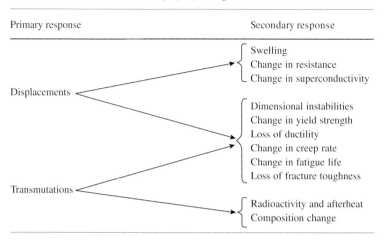

Primary response	Secondary response
Displacements	Swelling Change in resistance Change in superconductivity
	Dimensional instabilities Change in yield strength Loss of ductility Change in creep rate Change in fatigue life Loss of fracture toughness
Transmutations	Radioactivity and afterheat Composition change

9.4.2
Material Property Changes

The primary responses (displacements and transmutations) cause changes in the physical and mechanical properties of materials that are subjected to neutron bombardment. These secondary responses are indicated in Table 9.5.

The properties of structural materials are affected by neutron radiation in many different ways. Displacements produce interstitial atoms and lattice vacancies in pairs. The interstitials are highly mobile and insoluble, so that they tend to form loops, leaving an excess of vacancies in the lattice. These vacancies precipitate into three-dimensional voids if the temperature is sufficiently high and if there are nucleation sites. This microstructural change has no macroscopic effect until after a certain threshold dpa level has been reached, after which the material begins to swell. After the incubation dose has been exceeded, the volumetric swelling in structural materials increases linearly with the number of displacements per atom. The projected swelling rate for 316 SS, a Ti-modified version of 316 SS, and a ferritic steel are shown in Figure 9.9. The swelling characteristics are material and temperature dependent.

The helium produced in the material by (n, α) transmutation reactions can alter the microstructural evolution of irradiatiated materials, affecting the cavity formation and the precipitate and dislocation loop formation. Swelling in stainless steel is maximized at He/dpa ratios produced by high energy fusion neutrons. Helium bubbles on grain boundaries also can cause severe embrittlement at high temperatures.

Hardening of the lattice matrix in the grains of metals due to displacements causes deformation to occur along grain boundaries under stress. Helium collects on the grain boundaries, aggravating the problem by interfering with grain boundary sliding, which would otherwise relieve the stress. This process produces microcracks

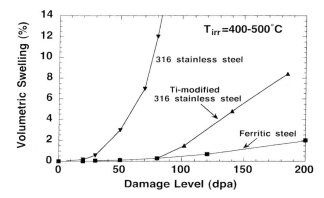

Figure 9.9 Volumetric swelling in steels under irradiation. (Reproduced with permission from ORNL Fusion Materials Program website.)

which eventually lead to fracture. One measure of ductility is the percentage of tensile elongation that is permissible without fracture. The effect of neutron radiation damage upon the tensile elongation of 316 stainless steel is depicted in Figure 9.10.

Defects caused by displacements can act as barriers to dislocation motion, causing metals to become stronger. The yield strength of some structural materials will generally increase under neutron irradiation by as much as a factor of $\sim$3. The effects of radiation on the yield strength of V-4Cr-4Ti are shown in Figure 9.11. This radiation

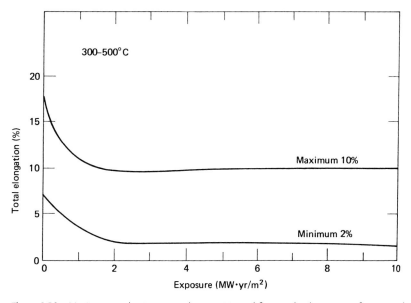

Figure 9.10 Maximum and minimum values anticipated for tensile elongation of 316 stainless steel as a function of fluence. (Reproduced with permission from W.M. Stacey et al., "US FED-INTOR", Ga. Inst. Techn report, 1982.)

**Effect of Dose and Irradiation Temperature on
the Yield Strength of V-(4-5%)Cr-(4-5%)Ti Alloys**

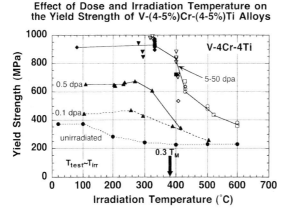

Figure 9.11 Hardening of yield strength under irradiation for V-4Cr-4Ti. (Reproduced with permission from ORNL Fusion Materials Program website.)

hardening induces an increase in the ductile to brittle transition temperature in body-centered-cubic metals.

Materials under thermal stress tend to deform slowly to relieve that stress, a process known as thermal creep. Thermal creep for V-4Cr-4Ti is shown in Figure 9.12. Neutron radiation has been observed to enhance the normal thermal creep rate. The additional "irradiation" creep is proportional to the number of displacements per atom.

When a structural material is repetitively cycled under a loading that produces a strain ε, it will eventually fail. The number of cycles to failure is directly related to the magnitude of the strain. When a material is irradiated with neutrons, this "fatigue" lifetime is decreased, as illustrated in Figure 9.13.

If a small flaw exists in a structural component, that flaw will grow under cyclic loading until fracture takes place. The fracture mechanism is plastic instability on a local scale. The resistance to flaw growth can be characterized by a parameter known as fracture toughness. Neutron radiation reduces the fracture toughness of structural materials.

While the existing structural materials irradiation database may be adequate, with some supplementation, for ITER, there is a large extrapolation to the database that will be needed for the design of future fusion reactors, as depicted schematically in Figure 9.14. While theory and modeling can guide the development of structural materials needed for the design of future reactors, and intense neutron source experimental facility will be needed to qualify the fusion structural materials used in these future reactors.

Superconducting magnet components are adversely affected by radiation, also. The effect of neutron radiation on the superconductor itself is to diminish the superconducting region of current density – temperature – magnetic field phase space. The critical current density of Nb_3Sn has been observed to deteriorate severely above $\sim 1.5 \times 10^{-3}$ dpa (3×10^{22} n/m^2). The effect is much less pronounced for NbTi,

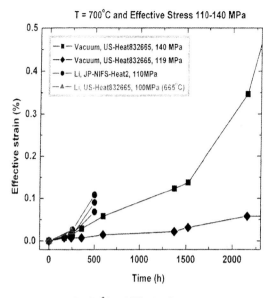

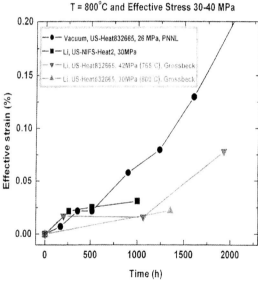

Figure 9.12 Irradiation-induced thermal creep in V-4C4-4Ti. (Reproduced with permission from ORNL Fusion Materials Program website.)

with a decrease of less than a factor of 2 being observed at fluences up to $10^{24}\, n/m^2$. These data are shown in Figure 9.15.

The resistance of normal conductors, which are used for cryogenic stability, at 4.2 K increases dramatically with the number of displacements per atom, as illustrated in Figure 9.16.

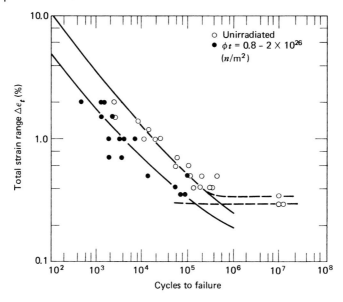

Figure 9.13 Fatigue life of 20% cold-worked 316 stainless steel irradiated at 430 °C and tested at irradiation temperature. (Reproduced with permission from W.M. Stacey *et al.*, "US INTOR", Ga. Inst. Techn. report 1981.)

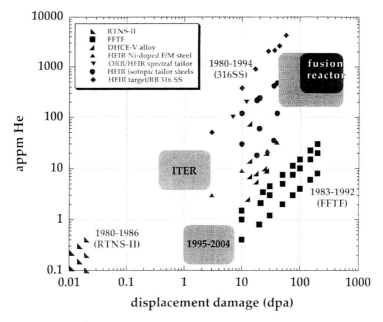

Figure 9.14 Helium production and damage (dpa) parameter space. (Reproduced with permission from ORNL Fusion Materials Program website.)

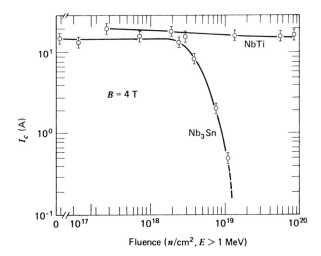

Figure 9.15 Effect of neutron irradiation on the critical current in NbTi and Nb₃Sn. (Reproduced with permission from "INTOR Phase Zero", IAEA report STI/PUB/556, Vienna, 1980.)

The organic insulators that are used in superconducting magnets have been observed to deteriorate physically from neutron damage in the range of 2–5 × $10^{22}\,n/m^2$. Inorganic insulators, on the other hand, retain their physical properties up to damage in excess of $10^{25}\,n/m^2$. However, the inorganic insulators are brittle.

There are indications that neutron radiation-induced microstructural change in granular solid breeding materials (see Section 10.1) may significantly reduce the tritium release properties. Tritium trapping at radiation-induced defects could be significant. Radiation-induced sintering could lead to reduction in porosity, which would reduce the tritium migration. The size of the basic grains could grow under radiation, leading to reduced tritium diffusion out of the grains in which it is produced.

9.5
Radioactivity

Many of the new isotopes produced by nuclear transmutations following neutron capture [e.g., (n, γ), (n, α), (n, p), $(n, 2n)$] are unstable and will, over a period of time, decay by the emission of a charged particle or a gamma ray. The resulting isotope may itself be unstable and undergo further decay, thus leading to a so-called decay chain. The nuclide densities of the isotopes in a given decay chain are described by the set of equations

$$\frac{dN_i(\mathbf{r}, t)}{dt} = -[\lambda_i + \bar{\sigma}_i \bar{\phi}(\mathbf{r})] N_i(\mathbf{r}, t)$$

$$+ \sum_{\substack{j \neq i}}^{J} [\lambda_j a_{j \to i} + \bar{\sigma}_{j \to i} \bar{\phi}(\mathbf{r})] N_j(\mathbf{r}, t) \quad i = 1, \dots, J$$

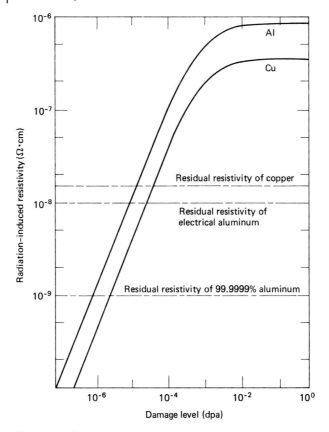

Figure 9.16 Radiation-induced resistivity of copper and aluminum. (Reproduced with permission from "INTOR Phase Zero", IAEA report STI/PUB/556, Vienna, 1980.)

Here λ_i is the radioactive decay constant for isotope i and $\alpha_{j\rightarrow i}$ is the probability that the decay of isotope j leads to isotope i. (For example, if the only decay path for isotope j is via electron emission, then $\alpha_{j\rightarrow i} = 1$ for isotope i with atomic number one greater and atomic mass the same, and $\alpha_{j\rightarrow i} = 0$ for all other isotopes i.) $\bar{\sigma}_i$ represents the spectrum-averaged microscopic cross section for all transmutation reactions [(n, α), (n, p), $(n, 2n)$, etc.], while $\bar{\sigma}_{j\rightarrow i}$ are the spectrum-averaged microscopic cross sections for the transmutation of isotope j into isotope i, if any exist. $\bar{\phi}$ is the integrated neutron scalar flux. A decay chain may have several branches corresponding to different decay processes (e.g., electron or proton emission) and different transmutation reactions. A given branch will terminate when a stable isotope is reached. There will be many such decay chains in any fusion reactor blanket or shield.

When a fusion reactor initially starts up, the concentration of the first isotope in the decay chain will grow until it reaches an equilibrium level where its decay and destruction by nuclear transmutation balances its production by nuclear transmutation. Then the second isotope in the decay chain will come into equilibrium, followed by the third, and so on until finally the entire chain reaches an equilibrium.

The characteristic time constant for each isotope is

$$\tau_i = (\lambda_i + \bar{\sigma}_i \bar{\phi})^{-1}$$

Once the isotopes above isotope i in the decay chain have reached equilibrium, then isotope i approaches equilibrium with a time dependence $\exp(-t/\tau_i)$.

The conventional measure of radioactivity is the curie (Ci), defined as

$$1 \text{ Ci} = 3.7 \times 10^{10} \text{ disintegrations per second}$$

The buildup of radioactivity in a representative fusion reactor blanket is illustrated in Figure 9.17 for five different choices of structural materials, 316 stainless steel, TZM (0.45% Ti, 0.1% Zr, 99.45% Mo), Nb-1Zr, V-20 Ti, and 2024 series aluminum alloy, which span the range of structural materials then under consideration.

The level of radioactivity builds up to an equilibrium value quite rapidly. The level is significant in all cases. For example, if the plant output was 2000 MW$_\text{th}$, the radioactive inventory would be $\sim 2 \times 10^9$ Ci for stainless steel, aluminum, and V-20 Ti to 4–5 $\times 10^9$ Ci for TZM and Nb-1Zr.

When the reactor is shut down, the production of radioisotopes via nuclear transmutation is terminated and the nuclide concentrations decrease. Note that the first isotope in a decay chain simply decays as $\exp(-\lambda_i t)$, whereas the decay of subsequent isotopes in the chain depends upon the decay of precursors.

The decay of radioactivity in a typical fusion reactor blanket following shutdown is depicted in Figure 9.18 for the same five structural materials.

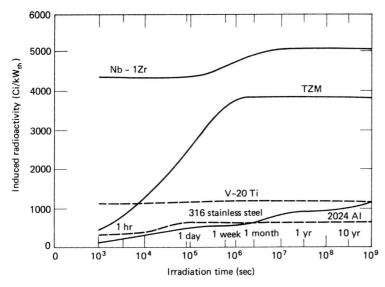

Figure 9.17 Buildup of blanket radioactivity following startup. (Reproduced with permission of Am. Nuc. Soc. from W.F. Vogelsang, CONF-760935, 1976.)

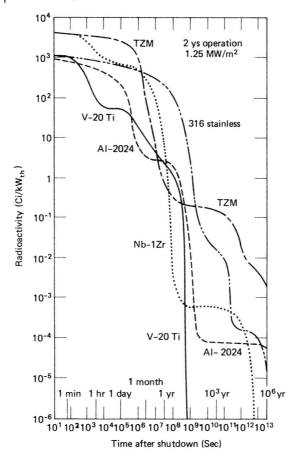

Figure 9.18 Fusion blanket radioactivity after shutdown. (Reproduced with permission of Am. Nuc. Soc. from W.F. Vogelsang, CONF-760935, 1976.)

The level of radioactivity would still be sufficiently high a few hours to a few days after shutdown of a 1000-MW$_{th}$ level fusion reactor that human access to the reactor internals would be precluded. Thus, radioactivity imposes a requirement for remote maintenance. Over the longer term, the radioactivity level is reduced by three to four orders of magnitude in 1–10 yr after which the subsequent decay tends to be determined by a single long-lived isotope in the decay chain. Efforts are presently in progress to "tailor" structural materials to minimize the isotopes which are precursors to these long-lived isotopes.

The radioactive decay produces heat. Cooling must be provided to remove this "afterheat," which decreases from about 1% of the reactor operating power immediately after shutdown to about 10^{-5}% 10 yr later.

9.6
Radiation Shielding

The purpose of radiation shielding is to attenuate the neutron and gamma radiation in order to limit radiation damage and nuclear heating of sensitive components (e.g., superconducting magnets) and to limit the activation of external reactor components to a level that will admit personnel access within a reasonable period after shutdown. The blanket, which directly surrounds the plasma chamber, acts as a shield also in removing about 99% of the neutron energy and in attenuating the incident neutron flux by a few orders of magnitude. This blanket, in turn, is surrounded by a bulk shield which attenuates the neutron and gamma fluxes to an acceptable level. Special shielding is also located around any penetrations (e.g., for neutral beam injectors) through the bulk shield and blanket and around any sensitive components. The wall of the reactor hall acts as a final "biological" shield to protect personnel in the control room during operation.

A few general criteria serve to guide the choice of materials for a radiation shield. A high atomic number material (e.g., tantalum, lead, tungsten, steel) is needed for the attenuation of gammas and high-energy neutrons. A light material (e.g., graphite, water, concrete) is needed to degrade the neutron energy, and a material with a large absorption cross section for low-energy neutrons (e.g., boron, lithium) is then needed to absorb the degraded neutrons.

Some representative shield compositions are given in Table 9.6. In each case the shield surrounds a blanket region. In some instances a highly scattering "reflector" region is interposed between blanket and shield to return some escaping neutrons to the blanket. Note that in designs $E - H$ the blanket region does not contain lithium and thus is really also a shield. The energy attenuation effectiveness of the various shields is compared in Figure 9.19. An energy attenuation of about 5 orders of magnitude is required in order to protect the superconducting magnets, which requires $\gtrsim 1$ m of blanket plus shield thickness.

The present fast neutron (> 0.1 MeV) limit on radiation to Nb_3Sn superconductor, $10^{19}\, n/cm^2$, and the dose limit to organic/inorganic insulation, $10^9/10^{10}$ rads, usually sets the shield design criterion.

Table 9.6 Several typical shield compositions[a].

Design	Blanket	Reflector	Shield
A	0.25 m, 90% Li + 10% V	0.05 m, SS	50% SS + 50% B$_4$C
B	0.25 m, 90% Li + 10% V	0.05 m, SS	75% SS + 25% B$_4$C
C	0.25 m, 90% Li + 10% V	0.05 m, SS	75% Pb + 25% B$_4$C
D	0.25 m, 90% Li + 10% V	0.05 m, SS	75% SS + 20% H$_2$O + 5% B
E	0.3 m, 50% SS + 50% B$_4$C		50% SS + 50% B$_4$C
F	0.3 m, SS		50% SS + 50% B$_4$C
G	0.3 m, 50% W + 50% B$_4$C		50% W + 50% B$_4$C
H	0.3 m, graphite		50% Al + 50% B$_4$C

a)   All percentages are by volume.

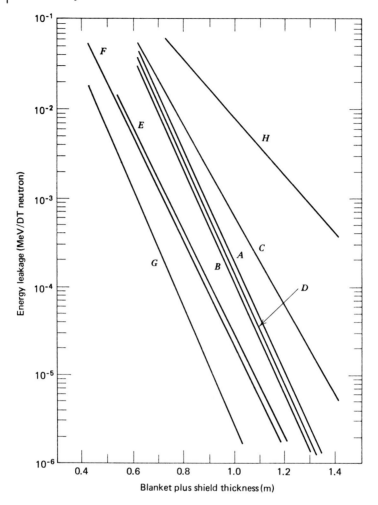

Figure 9.19 Energy leakage from the bulk shield as a function of the blanket/bulk shield thickness for several material compositions. Letters refer to the material compositions shown in Table 9.6. (Reproduced with permission from W.M. Stacey *et al.*, Argonne Nat. Lab. report ANL/CTR-75-2, 1975.)

Problems

9.1. Consider a tokamak reactor with $R = 5\,\mathrm{m}$ and $r_w = 1.3\,\mathrm{m}$ that produces $1000\,\mathrm{MW_{th}}$ power. Calculate the 14-MeV neutron current ($n/\mathrm{cm^2\,sec}$) that passes through the first wall.

9.2. Estimate the average volumetric nuclear heat deposition ($\mathrm{W/cm^3}$) for a 50-cm-thick annular blanket surrounding the plasma chamber in Problem 9.1.

9.3. Assume that the neutron flux attenuates exponentially into the blanket of the above problems with an *e*-folding distance of 15 cm and that the total microscopic displacement cross section for stainless steel, averaged over the neutron spectrum, is $\sigma_d = 15 \times 10^{-21}\,\text{cm}^2$. Calculate the dpa distribution within the blanket after 1-yr operation. Repeat for 5 yr.

9.4. Estimate the volumetric swelling of stainless steel 1 and 10 cm into the blanket for Problem 9.3 after 1- and 5-yr operation.

9.5. Consider a tokamak reactor with a neutron wall load of 2.5 MW/m². An annular blanket and shield 1.5 m thick surrounds the plasma chamber. Assume that the neutron flux is attenuated exponentially with an e-folding distance of 10 cm. How long can the reactor be operated before the superconducting magnet insulators reach their limiting neutron fluence of $2 \times 10^{22}\,n/\text{m}^2$? (The magnets are just outside the shield.)

9.6. Calculate the half-life (the time in which one-half the original concentration has decayed) after reactor shutdown for the decay of a radioactive isotope which has no radioactive precursors and a decay constant $\lambda = 0.01\,\text{hr}^{-1}$.

9.7. Consider a two-isotope decay chain with $\lambda_1 = 10^{-3}\,\text{hr}^{-1}$ and $\lambda_2 = 10^{-2}\,\text{hr}^{-1}$. While the reactor is running, the first isotope is produced by nuclear transmutation at a rate of $10^{25}\,\text{m}^{-3}\,\text{hr}^{-1}$. Neither isotope has a transmutation reaction with neutrons. Calculate the concentrations of both isotopes when the reactor runs for 6 months then shuts down for 2 months.

10
Primary Energy Conversion and Tritium Breeding Blanket

Immediately surrounding the plasma chamber is a region referred to as the blanket. The blanket has two principal functions; (i) to convert the fusion energy into heat, and (ii) to produce tritium. Of the energy produced by D-T fusion 80% leaves the plasma immediately in the form of kinetic energy of the 14-MeV neutrons. These neutrons transfer their energy to the blanket materials, as described in Section 9.3, constituting as such a volumetric heat source in the blanket. The remaining 20% of the energy produced in fusion and all energy externally supplied to heat the plasma ultimately leave the plasma as energetic charged particles or atoms or as radiative energy, in any case constituting a surface heat source incident on the first material surface facing the plasma. This first surface is also subjected to substantial particle fluxes, leading to erosion. Mechanisms employed to control the plasma edge region (e.g., limiters, divertors) tend to concentrate these particle and heat fluxes in localized areas. Special components are required to handle these concentrated heat and particle fluxes.

10.1
Tritium Breeding

The blanket surrounds the plasma chamber. It consists of a lithium-containing tritium breeding material, a coolant to remove the neutron-deposited heat, a structural material, and possibly other materials, such as neutron multipliers to enhance the tritium production. There are many possibilities for configuring the blanket, some of which are illustrated in Figure 10.1. The tritium breeding material may be solid or liquid. The blanket may be configured of modules of ~1-m dimension with the breeder in solid block or pool form and the coolant flowing in tubes (Figure 10.1a), or with the breeder contained in rods or tubes over which the coolant flows (Figure 10.1b). Alternatively, a liquid breeder may flow in vertical tube banks (Figure 10.1c), or a solid breeder may be configured in a monolithic cell that corresponds to some significant fraction of the entire blanket (Figure 10.1d). There are many variants of these basic configurations.

Fusion: Second, Completely Revised and Enlarged edition. Weston M. Stacey
Copyright © 2010 Wiley-VCH Verlag GmbH & Co. KGaA, Weinheim
ISBN: 978-3-527-40967-9

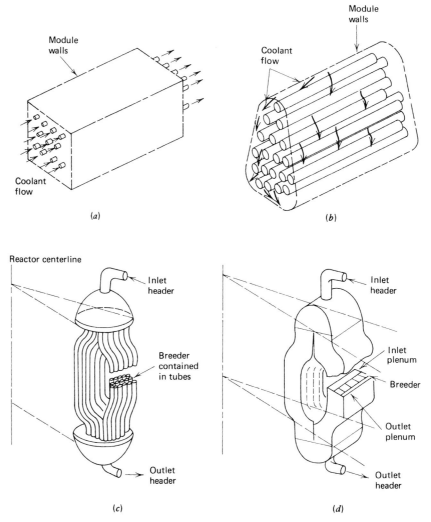

Figure 10.1 Blanket geometric and pressure containment concepts: (a) coolant contained within small tubes; (b) coolant contained within module walls; (c) vertical tube bank; (d) monolithic cell. (Reproduced with permission from C.C. Baker, Argonne National Lab. report AN L/FPP-80-1, 1980.)

The basic tritium breeding reactions are

$$n + {}^6\mathrm{Li} \rightarrow \mathrm{T} + {}^4\mathrm{He}$$

and

$$n + {}^7\mathrm{Li} \rightarrow \mathrm{T} + {}^4\mathrm{He} + n'$$

The first reaction has a large cross section, $\sigma = 950$ barns $= 9.5 \times 10^{-22}\,\mathrm{cm}^2$, for thermal neutrons, and the second reaction has a smaller cross section for fast

Table 10.1 Some tritium breeding materials.

Material	Melting point (°C)	Air/water reaction	Tritium breeding yield
Liquid			
Li	180	High[a]	Good
$Li_{17}Pb_{83}$	235	Moderate[a]	Very good
$Li_1Pb_4Bi_5$	100	Moderate[a]	Very good
$Li_{62}Pb_{38}$	464	High[a]	Very good
Solid			
Li_2O	1430	Low[b]	Good
$LiAlO_2$	1610	No	Fair
Li_2SiO_2	1200	No	Fair
Li_4SiO_4	1250	No	Fair
Li_2ZrO_3	1620	No	Fair
Li_7Pb_2	726	Yes	Good
LiAl	600	Moderate[a]	Fair
Li_3Bi	1145	Moderate[a]	Good

a) Release of H on reaction with H_2O.
b) Forms LiOH on contact with H_2O.

neutrons. Natural lithium contains 7.5% ^6Li and 92.5% ^7Li. It is possible to enrich the lithium to increase the ^6Li content.

The principal characteristics of some potential tritium breeding materials are listed in Table 10.1, and some properties of the materials of greatest current interest are given in Table 10.2. The liquid breeders, as a class, have the advantage that they can be circulated through the blanket, which allows the bred tritium to be extracted by processing the breeder external to the blanket. Tritium can be readily extracted from the liquid breeders; in fact there is some concern for LiPb alloys that it will come out of solution too readily while the breeder is still in the blanket, which could create a high

Table 10.2 Properties of some tritium breeding materials.

Material	Solid				Liquid	
	Li_2O	Li_2SiO_3	Li_4SiO_4	$LiAlO_2$	Li	$Li_{17}Pb_{83}$
Theoretical density (g/cm³)	2.01	2.53	2.39	α – 3.40 γ – 2.56	0.50	9.4
Molecular weight	29.88	89.96	119.84	65.92	6.94	NA
Lithium atom theoretical density (g/crn³)	0.93	0.39	0.55	0.27	0.50	0.056
Coefficient of thermal expansion (10^{-5}/K)	1.3	1.1	0.5–5.0	1.2	5.6	70
Heat capacity (J/g mol K)	54	100–150	140		4140[c]	170[c]
Thermal conductivity (W/m K)	4–6[a]	2–5.5[b]	2–5.5[b]	3.5–4.7	50	17

a) 90% theoretical density.
b) 60% theoretical density.
c) J/kg K.

partial pressure of tritium, possibly resulting in large tritium permeation rates. A difficulty with the liquid breeders is that their melting point is above room temperature, which requires that internal heaters must be provided to maintain the breeder molten during reactor shutdown or to melt it before startup. Liquid metals, particularly those containing lead, tend to be rather corrosive, which is another problem associated with the liquid breeders as a class.

Lithium undergoes exoergic reactions with air and water, which constitutes a potential safety problem. These reactions are suppressed when lithium is in the form of a binary or ternary ceramic compound (the oxygen-containing compounds in Table 10.1). The vigor of the reaction can also be reduced substantially by reducing the lithium density (e.g., $Li_{17}Pb_{83}$, which is in the ratio of 17 lithium atoms to 83 lead atoms). The heats of reaction of several breeding materials are given in Table 10.3.

The tritium breeding yield is generally improved with increasing lithium atom densities, with the exception of those breeding materials which themselves contain a good neutron multiplier such as Pb or Bi. The breeding materials indicated as "good" in Table 10.1 are capable of achieving a tritium breeding ratio (TBR) (i.e., number of tritium atoms produced per fusion neutron) in excess of unity, which provides for the possibility of a self-sustained D-T fuel cycle. A neutron multiplier must be included in the blanket in order to achieve TBR > 1 with those breeding materials indicated as "fair" in Table 10.1.

The achievable tritium breeding ratio depends on the thickness of the blanket. The results of a series of calculations for an idealized annular blanket surrounding the plasma region are given in Figure 10.2 for three different breeding materials. In the basic calculations the blanket was assumed to consist only of the breeding material, except that for the solid breeders (Li_2O and Li_7Pb_2) 15% of the breeding zone was assumed to be occupied by helium coolant. Auxiliary calculations were then made to assess the effect of (1) placing a 5-cm-thick beryllium neutron multiplying region between plasma and blanket, which improves the TBR significantly; (2) placing a 20-cm-thick graphite region behind the blanket to reflect escaping neutrons back into the blanket, which also improves the TBR; and (3) placing a 1-cm-thick vanadium first

Table 10.3 Heat of reaction[a] ΔH_R of breeders with water and air.

Breeder	ρ_{Li} (g/cm^3)	Reaction with water		Reaction with air (O$_2$)	
		(kJ/g Li)	(kJ/cm^3)	(kJ/g Li)	(kJ/cm^3)
Li	0.50	-34.3	-16.5	-25.9	-12.4
$Li_{17}Pb_{83}$	0.064	-28.4	-2.4	-20.0	-1.8
Li_2O	0.93	-8.4	-7.8	0.0	0.0
$LiAlO_2$	0.24	-0.9	-0.2	$+10.5$	$+1.8$
Li_2SiO_3	0.36	$+2.1$	$+0.8$	$+10.5$	$+3.8$

Reproduced with permission from W.M. Stacey *et al.*, "US INTOR," Georgia Institute of Technology report 1981.
a) $-$ exoergic, $+$ endoergic.

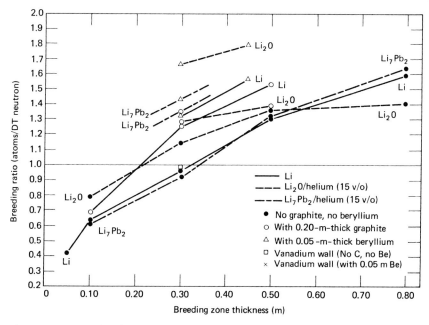

Figure 10.2 Tritium breeding ratio versus breeding zone thickness. (Reproduced with permission from D.L. Smith *et al.*, Argonne National Laboratory report ANL/FPP-79-1, 1979.)

wall between plasma and blanket, which substantially reduces the TBR because of neutron capture in the first wall.

The actual tritium breeding ratio that is achievable will be less than that shown in Figure 10.2 for two reasons. First, in a real reactor it will not be possible to occupy the entire region surrounding the plasma with a breeding blanket. A significant area ($\sim$10–20%) must be allocated to penetrations for heating systems, vacuum ducts, diagnostics, divertor channels, and so on, thereby reducing the TBRs of Figure 10.2 proportionally. Second, structural material will be present in an actual blanket.

The amount of structural material present in the breeder has an important effect on the achievable TBR. This is illustrated in Figure 10.3, where the deleterious effect of increasing parasitic neutron capture with increasing structure is apparent. This figure also shows the advantage of H_2O coolant relative to helium. This advantage results from the lower energy neutron spectrum due to neutron moderation by H_2O, leading to increased neutron capture in ^6Li, which more than offsets the reduction in ^7Li neutron capture and the parasitic neutron capture in H_2O.

The most probable form of a solid breeder in a blanket is illustrated in Figure 10.4. The breeding material will be in small grains ($<$1 µm) which are then formed into particles ($\sim$1 mm) with fine porosity. The particles, in turn, are packed into beds with a coarse porosity among particles. The packed beds are perforated with small-diameter ($\sim$2-mm) channels through which a low-pressure helium purge gas flows to recover tritium and carry it to an external processing system. The tritium produced within the grains must diffuse to the grain surface, desorb as T_2O, migrate through

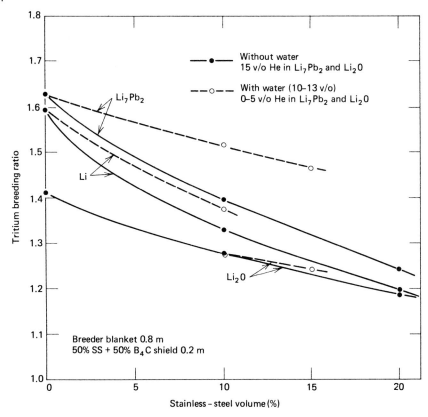

Figure 10.3 Impact of structural material (316 stainless steel) content on tritium breeding. (Reproduced with permission from D.L. Smith *et al.*, Argonne National Laboratory report ANL/FPP-79-I, 1979.)

the fine-grain-structure porosity and then "percolate" through the coarse-particle-structure porosity to the helium purge stream.

In situ recovery of tritium from a solid breeding blanket imposes limits on the operating temperature of the breeder. The migration rate of the bred tritium through the bilevel porosity structure (grains/particles) to the purge stream is not very temperature dependent, but the diffusion of the tritium out of the grains increases strongly with temperature. On the other hand, when the temperature exceeds ~80% of the melting temperature, restructuring and sintering of the grains may occur, which reduces the porosity and thereby decreases the migration rate. There is some evidence that neutron bombardment may also lead to sintering taking place at lower temperatures (~60% of the melting temperature). Thus, there are upper and lower temperature limits. A quantification of these limits may be specified by determining the temperature range over which the tritium removal rate is sufficiently large so that the tritium held up in the blanket is less than 1–2 kg for a few thousand thermal megawatt level reactor. Since the temperature will vary considerably over the blanket, as discussed in the next section, a temperature range of a few hundred degrees is a

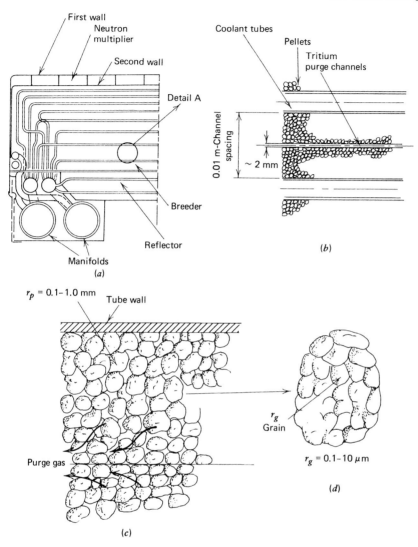

Figure 10.4 Schematic diagram of solid breeder blanket concept showing solid breeder microstructure with bimodal pore distribution and tritium removal scheme: (a) solid breeder; (b) section A–A; (c) packed bed; (d) pellet. (Reproduced with permission from W.M. Stacey et al., "US INTOR", Ga. Inst Techn. report, 1981.)

practical necessity, which eliminates certain of the potential solid breeders from further consideration.

The main line of ceramic breeder materials research and development is based on the use of the breeder material in the form of pebble beds. The pebbles are quasi-spheroid, with small dimensions (d < 1 mm), because they have a better margin against thermal cracking, can easily fit into complex blanket geometries, and better accommodate volumetric swelling and expansion. Three materials have evolved as

Table 10.4 Operating temperature windows for ceramic breeders.

Breeder	T_{min} (°C)	T_{max} (°C)
Li_4SiO_4	325	925
Li_2TiO_3	400	900
Li_2O	397	795

leading candidates for DEMO blankets that will be tested in ITER. These are Li_4SiO_4 pebbles produced by melt-spraying, Li_2TiO_3 pebbles produced by extrusion-spheronization-sintering, and Li_2TiO_3 pebbles produced by a wet process.

Some issues have not yet been completely resolved regarding the maximum allowable operating temperature at high fluence/high constraint for these candidate ceramic breeder materials. The current knowledge base includes tritium release data under low constraint conditions, and temperatures up to 700 °C. These data are at DEMO relevant burn-ups, that is, up to 11% for Li_4SiO_4 and 17% for Li_2TiO_3. In addition, from a thermochemical stability point of view, the spinel phase ($Li_4Ti_5O_{12}$) present in stoichiometric Li_2TiO_3 pebbles can react with H_2 in the purge gas and decompose into a completely different Li-oxide at high temperature. Tests for exploring high temperatures under a prototypical ratio of dpa/^{6}Li burn-up have already been planned in high fluence irradiation ceramic breeder pebble stacks tests (HICU) for all three reference materials.

The current best estimates of the allowable working temperature windows for the primary ceramic breeding materials are given in Table 10.4.

10.2
Blanket Coolants and Structural Materials

Four classes of coolants are being considered for fusion reactor blankets – gases, water, molten salts and liquid metals. Several types of structural materials have been considered – austenitic and ferritic stainless steels, high-nickel-based alloys, refractory alloys based on vanadium, niobium, molybdenum, and titanium, and more recently silicon carbide. The principal features of these coolants and structural materials are briefly reviewed.

Helium is the gas coolant with which there has been the most experience, primarily in the gas-cooled fission reactor field. Helium is an inert gas, which precludes many adverse chemical reactions that are a problem with other coolants. However, impurities in the helium present problems. It also has the potential for operating at very high temperatures to achieve high thermal-to-electrical conversion efficiencies. Helium, along with the other gases, has rather poor heat removal properties, which leads to a requirement for a large heat transfer area, large gas flow rates, and a corresponding large pumping power in order to cool the structure. Helium cooled reactors would ideally operate at coolant pressures of 5–10 MPa, maximum gas temperatures of 500–1000 °C, and gas flow velocities of 30–100 m/sec.

The advantage of high thermal-to-electrical energy conversion efficiency with helium can only be realized, of course, with high gas outlet temperatures, $\geq 400\,°C$. This requirement is in conflict with the poor heat transfer property of helium, which leads to a "film drop" temperature difference between the blanket structure and the gas of ~ 100–$250\,°C$, and the operating temperature limitations upon structural materials. Stainless steels suffer from embrittlement in a neutron environment due to transmutation reactions, which will probably limit the maximum operating temperature to $\lesssim 550\,°C$. Titanium alloys lose their strength above $500\,°C$, but these alloys are unacceptable in any case because of excessive levels of tritium absorption. High-nickel-base alloys can operate up to $650\,°C$ in a radiation environment before helium embrittlement becomes a limitation; unfortunately neutron radiation-induced embrittlement appears to preclude the use of such alloys. The refractory metals have good strength at elevated temperatures in a pure helium environment. However, there is evidence that trace impurities (e.g., H_2, H_2O, CO, CO_2, CH_4, N_2, O_2) in commercial-grade helium can lead to corrosion-embrittlement of vanadium, niobium, and molybdenum alloys – thereby rendering them unacceptable for use with a helium coolant. Thus, it appears that of the metallic structural materials, helium coolant can only be used with stainless steel and that the maximum steel temperature of $\lesssim 500$–$550\,°C$ probably limits the helium outlet temperature to $\lesssim 450\,°C$, which limits its potential high energy conversion efficiency. The recent development of SiC structural material provides the greatest promise for realizing the high temperature potential of helium.

Pressurized water (H_2O or D_2O) is an attractive coolant because of its good heat removal properties and the large base of experience from the fission reactor field. The principal limitations of water as a coolant relate to the requirements for high pressure and relatively low outlet temperature, to its chemical reactions with some of the potential tritium breeding materials, and to the difficulty of extracting any tritium which leaks into the cooling system. Pressurized water fission reactors operate with pressures of ~ 14 MPa ($\sim 2000\,\mathrm{lb/in^2}$) and maximum coolant temperatures of $\sim 320\,°C$, which limits the thermal conversion efficiency to $\lesssim 36\%$.

Water reacts exothermically with lithium-containing materials. The energy release in these and other hydrolysis reactions with breeding materials are given in Table 10.3. The compound Li_2O reacts much less exothermically, thus constituting much less of a hazard. The reactions for some of the candidate breeding materials are

$$Li + H_2O \rightarrow LiOH + \frac{1}{2}H_2$$

$$\frac{1}{7}Li_7Pb_2 + H_2O \rightarrow LiOH + \frac{1}{2}H_2 + \frac{2}{7}Pb$$

$$\frac{1}{2}Li_2O + \frac{1}{2}H_2O \rightarrow LiOH$$

The LiOH formed in these hydrolysis reactions has a melting point of $470\,°C$. Liquid LiOH is extremely corrosive to nearly all structural materials.

Stainless steel is compatible for use as a structural material with water, and the temperature limit on stainless steel ($\lesssim 450\,°C$) is consistent with the maximum water temperature (∼320 °C). Vanadium- and niobium-based alloys may be subject to substantial levels of corrosion in water. Hydrogen embrittlement of titanium alloys renders them unacceptable for use with water The high-nickel-base alloys are acceptable with water, but the radiation embrittlement probably precludes their use.

Lithium and other alkali liquid metals have excellent heat removal properties at low coolant pressures. With lithium or a lithium eutectic there exists the possibility that the coolant can also serve as the tritium breeding material. There is a significant base of liquid metal technology from the fast fission reactor program. The principal limitations relative to the use of lithium or a lithium-lead eutectic as a coolant are related to (1) corrosion of the structural material; (2) magnetohydrodynamic (MHD) effects on pressure drop and heat transfer; (3) high melting point; and (4) chemical reactivity with water and air.

Corrosion of structural materials is a major consideration for lithium-based coolants because of the deterioration of the mechanical integrity and because of the transport of radioactive corrosion products that might plug the system or deposit in unshielded regions. Austenitic stainless steels will probably be limited to ≲400–450 °C operating temperatures by corrosion effects in lithium. The ferritic steels are more corrosion resistant. The high-nickel alloys are not suitable in a lithium environment because of the high solubility of nickel in lithium, which leads to unacceptable corrosion. Alloying elements with titanium (e.g., tin and aluminum) are highly soluble in lithium, which probably renders titanium alloys unsuitable. Vanadium and niobium alloys have good corrosion resistance to lithium at temperatures more than 800 °C and are certainly the preferred structural materials for a liquid metal cooled blanket.

Some properties of the candidate coolants which determine their heat removal capabilities are given in Table 10.5. Properties of several candidate structural materials are given in Table 10.6.

Of the structural materials under consideration, there is a substantial basis of fabrication experience with stainless steel and titanium alloys from the nuclear and aerospace industries, respectively. The interest in structural materials other than austenitic stainless steels is motivated by their greater strength, superior thermophysical properties, or superior resistance to degradation of mechanical properties under neutron irradiation.

Neutron radiation affects the mechanical properties of structural materials, as discussed in Section 9.4. For austenitic stainless steels a substantial data base for the

Table 10.5 Some properties of potential coolants.

Coolant	Density ρ (kg/m^3)	Heat Capacity C_p (J/kg. K)	Thermal conductivity κ (W/m K)	Viscosity μ (Pa-sec)	Electrical conductivity (siemens/m)
He(7 MPa, 400 °C)	8.3	5.2×10^3	0.2	2.5×10^{-5}	—
H$_2$O(14 MPa, 300 °C)	950	4.5×10^3	0.5	9.5×10^{-4}	—
Li(0.1 MPa, 400 °C)	520	4.2×10^3	50	4.5×10^{-4}	1.5×10^6

Table 10.6 Mechanical and thermal unirradiated properties of several candidate structure materials for fusion.

Property	Austenitic stainless Steel (316 SS)	Ferritic stainless steel (HT-9)	High-nickel alloy (Inconel 625)	Titanium alloy (Ti-6242)	Vanadium alloy (V-15Cr-5Ti)	Niobium alloy (FS-85)
Melting Temperature (°C)	1400	1420	1400	1650	1900	2470
Density (kg/m³)	8×10^3	7.8×10^3	8.4×10^3	4.5×10^3	6.1×10^3	10.6×10^3
Poisson's ratio	0.27	0.27	0.28	0.32	0.36	0.38
Modulus of elasticity (GPa)						
400 °C	170	185	195	100	120	140
600 °C	145	160	170	70	115	135
Thermal expansion coefficient ($10^{-6}\,\mathrm{K}^{-1}$)						
400 °C	17.6	11.6	13.2	9.4	10.0	6.8
600 °C	18.4	12.2	14.0		10.4	7.2
Thermal conductivity (W/m K)						
400 °C	20	28	14	9	30	48
600 °C	22	27	18	11	35	50
Heat capacity (J/kg K)						
400 °C	540	700	500	650	530	230
600 °C	570	800	530	800	560	250
Yield strength (MPa)						
400 °C	120	375	400	600	500	300
600 °C	100	100	380	475	450	290
Ultimate tensile strength (MPa)						
400 °C	450	590	850	725	700	450
600 °C	375	200	800	600	650	400
Electrical resistivity (μΩ cm)						
400 °C	105	90	135	200	40	30
600 °C	110	110	140	200	50	35
Creep rupture stress limit (MPa)						
10^2 hr	400	700	700	500	450	
10^4 hr	200	500	500	300	400	

Table 10.7 Estimated swelling and ductility of early candidate alloys after 10 MW. yr/m^2 neutron irradiation.

Material	Peak Swelling		Ductility at 500–600 °C	
	Temperature (°C)	$\Delta V/V$ (%)	Uniform elongation (%)	Tensile elongation (%)
316SS 20% CW[a]	575	30	0–0.5	<1
Inconel 625	450	2	<0.1	<1
HT-9	400	1	1[b]	3
Ti-6242	—	—	—	—
V-15Cr-5Ti	600	<1	3	4
FS-85	800	6	0–1	<2

Reproduced with permission from D.L. Smith *et al.*, Argonne National Lab report ANL/FPP-29-1, 1979.
a) Cold-worked.
b) Increase in ductile–brittle transition temperature to above room temperature.

effects of radiation on materials properties exists. Estimated peak swelling and residual ductility after irradiation to 10 MW yr/m^2 (1.4×10^{25} n/m^2) of 14-MeV neutron fluence are presented in Table 10.7. The amount of swelling and the minimum ductility that are tolerable depend upon the design, but $\Delta V/V \lesssim 4\%$ and $\Delta l/l \gtrsim 1$–2% are representative numbers. A small amount of titanium added to an austenitic stainless steel can dramatically reduce the amount of swelling. An austenitic stainless steel with a composition tailored to minimize swelling was examined as a primary candidate alloy (PCA). The amount of swelling can also be dramatically reduced by operating at a reduced temperature for annealed (SA) and cold-worked (CW) 316 stainless steel.

The ferritic steels have superior swelling resistance properties to the austenitic steels. However, the transition temperature between a ductile and a brittle phase increases above room temperature under neutron irradiation. The high-nickel alloy suffers from radiation embrittlement. From the standpoint of radiation resistance, the vanadium alloy appears to be preferable, although the data are limited.

The activation characteristics of materials have an important influence on waste material disposal, the prospects for limited hands-on maintenance, and reactor safety. To this extent, lower activation materials, including vanadium and titanium alloys and ceramics such as silicon carbide and graphite, are favored. Steel alloys can be improved with respect to activation characteristics by carefully tailoring the elements in an alloy. In particular, highly activating elements such as nickel, molybdenum, and niobium should be avoided. Certain ferritic steels and high-manganese stainless steels are two general examples.

A silicon-carbide/silicon-carbide (SiC/SiC) composite has recently received more emphasis as a potential structural material for fusion. SiC has several attractive features, including high thermal conductivity and low neutron activation, but it is relatively less developed than the other materials under consideration for fusion reactors.

Table 10.8 Neutronic responses of first-wall/blanket structural materials[a].

Alloy	dpa[b] (15 MWy/m^2)	He transmut[b] appm (15 MWy/m^2)	H transmut[b] appm (15 MWy/m^2)	Dose rate[c] Sv/hr	Decay heat[c] W/kg
Austenitic steel	170	2400	8550	4000	3
Ferritic steel	170	1800	7350	1000	1
Vanadium alloy	170	855	4050	0.3	0.005
SiC/SiC comp	135	19 500	13 350	0.0001	0.00 003
Niobium alloy	95	495	1725	4000	4
Molybdenum alloy	95	525	5250	500	0.3
Tantalum alloy	51	45	135	1 000 000	1000
Tungsten alloy	45	51	135	1000	10
Copper	210	1500	8700	2000	1

a) Approximate values, actual values depend on specific design.
b) Approximate values at first wall after 15 MW y/m^2.
c) Approximate values 1 month after 3 Yr @ 5 MW/m^2.

A comparison of the calculated responses of the various candidate structural materials in a fusion reactor first wall exposed to a 14 MeV neutron flux of 5 MW/m^2 for 3 years is given in Table 10.8. Note 1) the very high H and He transmutation rate for the SiC/SiC composite and the very low H and He transmutation rate for the tantalum and tungsten alloys; 2) the dose rate and decay heat which are very high for tantulum and very low for the SiC/SiC and vanadium. Use of a tantalum alloy would appear to be unacceptable, and the high helium generation rate raises a major feasibility concern about SiC/SiC.

Based on the radiation damage data to date, the estimated allowable operating temperature ranges of the various potential fusion reactor structure materials are depicted in Figure 10.5. The 300 series stainless steels are constrained by thermal creep and poor swelling resistance at high burnups. The ferritic steels have excellent swelling resistance at high burnups, but are constrained by radiation hardening at low temperature and thermal creep at high temperatures. The "superalloys" (e.g., Inconel) avoid thermal creep at high temperature but are susceptible to radiation embrittlement at moderate neutron doses. The refractory alloys (V, Nb, Ta, Mo) have adequate swelling resistance up to high burnups, but their use is hampered by fabrication and joining difficulties, low temperature radiation hardening and poor resistance to oxidation.

Austenic stainless steels are limited by high swelling rates and low thermal conductivity, which has spurred the investigation of ferretic/martensitic (F/M) steels with better radiation resistance and thermal properties. The ductile-to-brittle transition temperature of early candidate F/M steels were found to rise above room temperature with neutron irradiation, but there has been good progress in tailoring the isotopic composition of F/M steels to eliminate this problem. However, if "conventional" high-chromium ferritic/martensitic steels were used as structural materials, the upper operating temperature for the first-wall or blanket structural

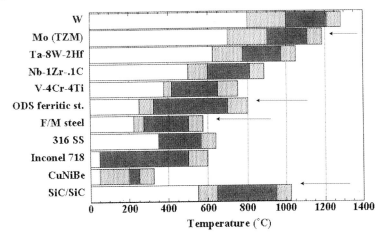

Figure 10.5 Structural Materials Operating Temperature Ranges. Darker areas indicate safe operating ranges, and shaded areas indicate marginal operating ranges. (Reproduced with permission from ORNL Fusion Materials Program website).

material of a fusion reactor would be limited to 550–600 °C. One possibility to increase this upper temperature limit, while maintaining the inherent advantages (higher thermal conductivity and resistance to radiation swelling), is to use oxide dispersion-strengthened (ODS) steels in which a high density of small Y_2O_3 or TiO_2 particles are dispersed in a ferrite matrix to achieve improved high temperature strength.

All of the refractory (W, Ta, Mo, Nb, V) alloys have good high temperature strength properties, but all with the exception of V-alloys become highly radioactive under neutron irradiation.

For these reasons, the primary candidate structural materials presently under development for future fusion reactors are the (i) F/M steels, particularly the ODS steels, and (ii) vanadium alloys as typified by V-4Cr-4Ti. Although not as developed as the F/M steels and V-alloys, (iii) SiC/SiC fiber composites have very good high temperature properties and are also being carried along as a primary candidate structural material for this reason.

The "design operating windows" for candidate ODS steel, V alloy and SiC/SiC structural materials are shown in Figures 10.6–10.8. The limiting phenomena in different regions are indicated.

Copper alloys and austenitic stainless steels are proposed for near-term applications, such as ITER, which will not be subjected to large neutron fluences.

Perhaps the most extensive of the fusion reactor structural materials development programs is that for an oxide dispersion strengthened (ODS) steel alloy. There are two main options for ODS steels: (i) ferritic ODS steels (~12–16% Cr), and (ii) ferritic/martensitic ODS steels (~9% Cr). The former has a higher temperature capability and better oxidation resistance, but lower fracture toughness, than the latter.

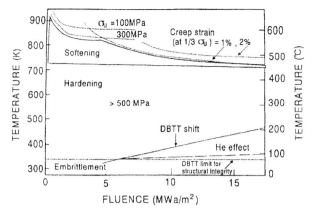

Figure 10.6 Operating limits for ferritic/martensitic ODS steel, with limiting phenomena. (Reproduced with permission from ORNL Fusion Materials Program website).

As discussed above, the primary candidate structural materials for advanced fusion systems are F/M steels, vanadium alloys and SiC/SiC composites. While it is not yet possible to predict with confidence the lifetime of these evolving advanced structural materials against radiation damage, based on what is now known it is not unreasonable to anticipate that radiation damage lifetimes of >200 dpa can be achieved in the future. F/M steel has been irradiated to 200 dpa without significant radiation damage.

The operating temperature and thermal conversion efficiency ranges of the future fusion reactor blankets that could be designed on the basis of the development of these advanced structural materials are depicted in Figure 10.9.

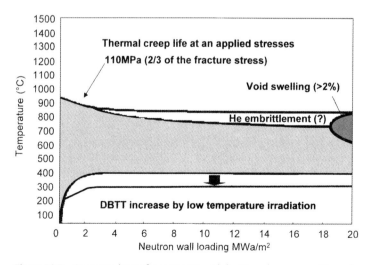

Figure 10.7 Operating limits for V-4Cr-4Ti, with limiting phenomena. (Reproduced with permission from ORNL Fusion Materials Program website).

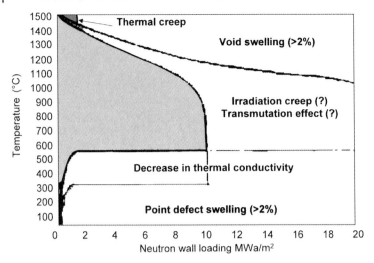

Figure 10.8 Operating limits for SiC/SiC, with limiting phenomena. (Reproduced with permission from ORNL Fusion Materials Program website).

10.3
Heat Removal, Hydraulics, and Stress

Heat transfer within the blanket region is described by the equation

$$\rho C_p \frac{\partial T}{\partial t} = \kappa \nabla^2 T + q'''$$

where ρ (kg/m^3) is the density, C_p (J/kg K) is the heat capacity, T (K) is the temperature, κ (W/m K) is the thermal conductivity of the blanket material, and

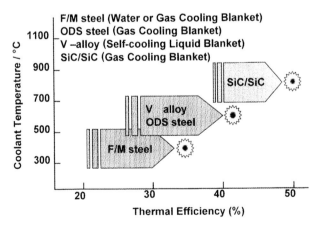

Figure 10.9 Future fusion reactor blanket temperature and thermal conversion efficiency ranges achievable with existing and advanced structural materials. (Reproduced with permission from ORNL Fusion Materials Program website).

q''' (J/m^3 s) is the volumetric nuclear heating source described in Section 9.3. At an interface between a stationary, usually solid, region and a flowing coolant (gas or liquid), the heat transfer is governed by

$$q'' = h(T_s - T_f)$$

where q'' (W/m^2) is the heat flux flowing across the interface, h (W/m^2 K) is the convective heat transfer coefficient, and T_s and T_f are the solid and fluid temperatures, respectively, on either side of the interface.

The convective heat transfer coefficient is determined from

$$h = \frac{\kappa(Nu)}{D}$$

where the Nusselt number Nu is correlated to the coolant properties as follows.

For gas (He) and liquid (H$_2$O),

$$Nu = 0.023 \, (Re)^{0.8}(Pr)^{0.4}$$

and for liquid metal,

$$Nu = \begin{cases} 7 + 0.025 \, (RePr)^{0.8}, & \text{constant heat flux along channel} \\ 5 + 0.025 \, (RePr)^{0.8}, & \text{constant wall temperature along channel} \end{cases}$$

where the Reynolds number

$$Re \equiv \frac{DV\rho}{\mu}$$

and the Prandtl number

$$Pr \equiv \frac{C_p\mu}{\kappa}$$

are defined in terms of the fluid viscosity μ (Pa sec), density ρ (kg/m^3), flow velocity V (m/sec), thermal conductivity κ (W/m·K), and tube diameter D (m). For noncircular coolant channels,

$$D = \frac{4 \times \text{flow area}}{\text{wetted perimeter}}$$

As a representative example, consider the solid breeder blanket configuration depicted in Figure 10.10. Coolant tubes are embedded in a monolithic block of solid breeding material. For a nearly uniform nuclear heating distribution, each tube removes the heat deposited in an annular region of radius R_3. The coolant tube is separated from the solid breeder by an insulating gap to prevent corrosion of the tube wall and to provide a means of controlling the temperature in the breeder by controlling the heat transfer across the gap. The gap may contain helium or another gas or a solid thermal insulating material.

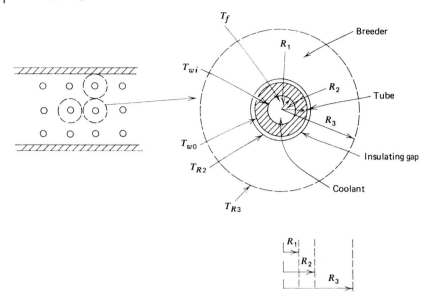

Figure 10.10 Solid breeder/coolant tube array.

The boundary conditions appropriate to this problem are

$$R = R_3 : \quad \frac{dT}{dr} = 0 \qquad\qquad \text{(symmetry)}$$

$$R = R_2 : \quad -\kappa_b \frac{dT}{dr} = h_g(T_{R2} - T_{wo}) \quad \text{(inner blanket-gap)}$$

$$R = R_1 : \quad -\kappa_s \frac{dT}{dr} = h_f(T_{wi} - T_f) \quad \text{(inner tube wall-coolant)}$$

where h_g is the convective heat transfer coefficient across the gap; h_f is the convective heat transfer coefficient between the coolant tube and the coolant; T_{R3}, T_{R2}, T_{wo}, T_{w1}, and T_f are the breeder temperatures at R_3 and R_2, the tube temperatures at R_2 and R_1, and the coolant temperature, respectively, and κ_b and κ_s refer to the thermal conductivities of the breeder and the coolant tube (structure), respectively.

The steady-state solution for the radial temperature distribution in the breeder region is

$$T(r) = T_{max} - \frac{q_b''' R_3^2}{4\kappa_b} \left[\left(\frac{r}{R_3}\right)^2 - 2\ln\left(\frac{r}{R_3}\right) - 1 \right]$$

where q_b''' is the nuclear heating in the breeder, and the maximum temperature is related to the coolant temperature T_f,

$$T_{R3} = T_{max} = T_f + \frac{q_b''' R_3^2}{4\kappa_b} \left[\left(\frac{R_2}{R_3}\right)^2 - 2\ln\left(\frac{R_2}{R_3}\right) - 1 \right]$$

$$+ \frac{q_b''' R_3^2}{2} \left[1 - \left(\frac{R_2}{R_3} \right)^2 \right] \left[\frac{1}{\kappa_s} \ln \left(\frac{R_2}{R_1} \right) + \frac{1}{R_1 h_f} \right]$$

$$+ \frac{q_b'''}{2 R_2 h_g} \left[R_3^2 - R_2^2 \right]$$

The coolant temperature increases along the coolant channel because of the heat input. In the steady state the temperature increase over an incremental length $\Delta L = L_{i+1} - L_i$ satisfies.

$$\rho A_c V C_p (T_f^{i+1} - T_f^i) \equiv \dot{W} C_p (T_f^{i+1} - T_f^i) = Q_i$$

where ρ, A_c, and V are the coolant density, flow area, and velocity; Q_i is the total heat input to the channel between L_i and L_{i+1} (Figure 10.11)

$$Q_i = \pi \Delta L \{ R_1^2 q_f''' + (R_2^2 - R_1^2) q_s''' + (R_3^2 - R_2^2) q_b''' \}$$

and q_f''' and q_s''' are the nuclear heat inputs to the coolant and the tube, respectively.

As the coolant temperature increases along the channel, the radial temperature distribution in the breeder increases correspondingly, as indicated in Figure 10.12. The breeding blanket must be designed so that the maximum and minimum temperatures at any point in the breeding material are within the allowable temperature range given in Table 10.4.

The pressure drop experienced by a coolant in flowing through a channel of length L_c is

$$\Delta p(\text{Pa}) = \frac{f L_c \rho V^2}{2D} + \Delta p_{\text{MHD}}$$

where f is the Moody friction factor, given in Figure 10.13, and

$$\Delta p_{\text{MHD}}(\text{Pa}) \simeq L_c V B_\perp^2 \sigma_f \frac{C}{1+C}$$

is the MHD pressure drop for a conducting (liquid metal) coolant. Here $B_\perp(T)$ is the component of the magnetic field perpendicular to the coolant flow direction, σ_f is the electrical conductivity of the coolant, and

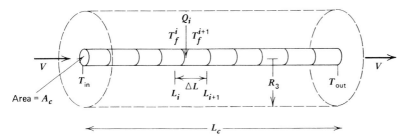

Figure 10.11 Coolant channel model.

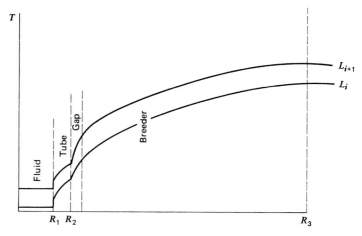

Figure 10.12 Radial temperature distributions.

$$C = \frac{2\sigma_s t}{\sigma_f D}$$

where σ_s and $t = R_2 - R_1$ are the electrical conductivity and the thickness, respectively, of the coolant tube wall.

The pumping power required to move the coolant through a channel against the pressure drop Δp is

$$P_{pc} = \frac{(\Delta p)\dot{W}}{\rho \eta_p} = \frac{(\Delta p)A_c V_c}{\eta_p}$$

where η_p is the efficiency of the pump. If the channel flow area A_c is replaced by the entire flow area of all channels, the above expression yields the total pumping power P_p.

The total thermal power removed from the blanket is

$$P_{th} = \dot{W}C_p(T_{out} - T_{in})$$

where T_{in} and T_{out} are the inlet and outlet coolant temperatures. Thus, the ratio of pumping power to thermal power may be written

$$\frac{P_p}{P_{th}} = \frac{\Delta p}{\eta_p C_p(T_{out} - T_{in})}$$

Representative flow rates and pumping powers for cooling a 1000-MW$_{th}$ fusion reactor blanket are given in Table 10.9 for the three different coolant options.

Stresses will be induced in the coolant tubes, primarily by the coolant pressure and the thermal gradients. For a coolant pressure p the tensile "hoop" stress in a coolant tube of radius R_1 and thickness $t = R_2 - R_1 \ll R_1$ is

$$\sigma_p = \frac{pR_1}{t}$$

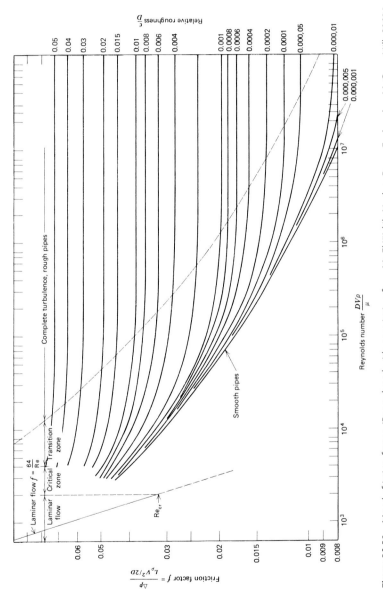

Figure 10.13 Moody friction factor. (Reproduced with permission from M. El-Wakil, *Nuclear Power Engineering*, McGraw-Hill, 1962.)

Table 10.9 Flow rates and pumping power for cooling a 1000-MW$_{th}$ fusion reactor blanket.

Coolant	$\dot{W}$ (kg/sec)	A_cV_c (m^3/sec)	P_p (MW)
He	1920	540	15
H$_2$O	2430	2.7	< 1
Li	2430	5.4	43.6 (43 MHD)

Reproduced with permission from T.J. Dolan, *Fusion Research*, Vol. III, Pergamon Press, 1982.

The thermal stress is

$$\sigma_{th} = \varepsilon_{th} E = \left[\frac{\alpha \Delta T}{2(1-\nu)} \right] E$$

where ε_{th} is the thermal strain associated with a temperature gradient ΔT across the tube wall, α is the coefficient of thermal expansion, ν is Poisson's ratio, and E is Young's modulus. For $t \ll R_1$,

$$\Delta T \simeq \frac{q''t + \frac{1}{2}q_s'''t^2}{\kappa_s}$$

where q'' is the heat flux into the coolant tube from the breeder region and κ_s is the tube thermal conductivity. Combining,

$$\sigma_{th} = \frac{\alpha E}{2\kappa_s(1-\nu)} \left(q''t + \frac{1}{2}q_s'''t^2 \right)$$

The thermal and hoop stresses, and their sum

$$\sigma = \sigma_{th} + \sigma_p$$

are plotted for a representative tube in Figure 10.14. The tube wall thickness must then be determined by applying the ASME[1] code, which defines allowable limits on σ_p and on $\sigma_{th} + \sigma_p$ in terms of the yield strength, ultimate strength, and creep properties of the material. The ASME criteria may be expressed in this case as

$$\sigma_p \leq S_{mt}(t)$$
$$\sigma_p + \sigma_{th} \leq \begin{cases} 1.5 S_m(t) \\ K_t S_t(t) \end{cases}$$

where $S_{mt}(t)$, $S_m(t)$, $S_t(t)$, and $K_t(t)$ are time-dependent material-dependent parameters. S_{mt} is the lower of S_m or S_t. The quantity S_m is defined as the lowest of one-third the ultimate strength S_u at room temperature or operating temperature or two-thirds the yield stress S_y at room temperature or operating temperature (0.9 S_y at operating temperature for austenitic stainless steel). The quantity $S_t(t)$ is the lowest of two-thirds the minimum stress to cause creep rupture in time t, 80% of the minimum stress to cause the onset of tertiary creep in time t, or the minimum stress to produce

---

1) American Society of Mechanical Engineers.

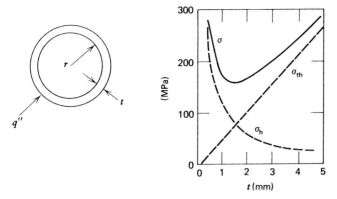

Figure 10.14 Thermal stress, hoop stress, and total stress versus tube thickness for a 3 16 stainless steel tube with $r = 2$ cm, $q'' = 0.5$ MW/m², $T = 500\,°$C, containing helium at 6 MPa, (Reproduced with permission from T.J. Dolan, *Fusion Research*, Vol. III. Pergamon Press, Elmsford, N.Y., 1982.)

1% total strain in time t. The quantity K_t is

$$K_t = 1 + \frac{1}{2}\left(1 - \frac{\sigma_p}{S_t}\right).$$

The thermal stress parameter

$$M \equiv \frac{\alpha E}{(1-\nu)\kappa}$$

provides a relative measure of the consequences of the thermophysical properties of the different candidate structural materials. This parameter is given in Table 10.10. The smaller M, the smaller the thermal stress for a given heat flux and tube thickness.

10.4
Plasma Facing Components

Those components whose surface is directly exposed to the plasma are subjected to high particle and heat fluxes. One-fifth of the energy produced in a D-T fusion event

Table 10.10 Thermal stress factors for candidate structural materials at probable mean operating temperatures.

Reference structural material	Assumed operating temperature (°C)	Thermal stress factor M (MPa m/W)
316 stainless steel	400	0.220
Inconel 625	500	0.218
Ti-6242	400	0.128
Fe-9Cr-1Mo	400	0.112
V-15Cr-5Ti	500	0.055
Nb (FS-85)	500	0.028

Reproduced with permission from D.L. Smith *et al.*, Argonne National Lab report ANL/FPP-79-1, 1979.

remains in the plasma and ultimately emanates from the plasma in the form of energetic particles or electromagnetic radiation. The average energy flux incident upon the wall surrounding a plasma that produces thermal power P_{th} is

$$\hat{q}''_w = \frac{1}{5}\frac{P_{th}}{A_w}$$

where A_w is the area of the wall. When a divertor or limiter is present, the charged particle fraction $f_{\pm}$ of this is concentrated upon the smaller area $A_{D/L}$ of the divertor collector plate or limiter, so that the heat fluxes to the wall and the divertor or limiter are

$$q''_w = \frac{1}{5}\frac{P_{th}(1-f\pm)}{A_w}$$

and

$$q''_{D/L} = \frac{1}{5}\frac{P_{th}f\pm}{A_{D/L}}$$

respectively. For tokamaks $A_{D/L} \simeq 0.1 A_w$, so that the divertor or limiter serves to concentrate the charged particle component of the heat flux by an order of magnitude. There is the possibility of increasing $A_{D/L}$ with certain types of divertor designs, and the tandem mirror configuration seems to lend itself to this particularly well. In principle $f_{\pm}$ can vary from near unity for a plasma with no impurities to near zero for a plasma with a radiating edge (see Section 6.2).

Cooling the special surface components (first wall, divertor collector plate, limiter) against the surface heat flux from the plasma is a prime consideration in the design of these components. Coolant channels are placed as close as possible to the surface. Many possible configurations are under consideration. For our purposes it suffices to consider the tube bank configuration shown in Figure 10.15. This general configuration applies both to the first wall of the blanket ($q''_p = q''_w$) and to the divertor or limiter ($q''_p = q''_{D/L}$).

The thermal stress resulting from a temperature gradient, ΔT, across the tube wall is

$$\sigma_{th} = E\varepsilon_{th} = \frac{E\alpha}{2\kappa(1-\nu)}q''_p\delta$$

The hoop stress due to the coolant pressure is

$$\sigma_p = \frac{pR}{\delta}$$

The minimum tube wall thickness is determined from the ASME code and from erosion considerations,

$$\delta_{min}(L_p) = \frac{pR}{S_{mt}(L_p)} + \delta_{eros}(L_p)$$

The first term gives the minimum thickness required to withstand the coolant pressure for a lifetime L_p. The second term gives the additional thickness that would be eroded away over the lifetime L_p (see Section 6.1).

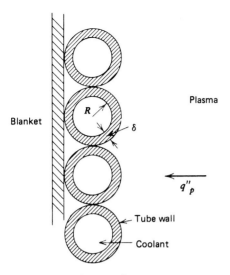

Figure 10.15 Plasma surface components.

For pulsed operation, fatigue lifetime limits the allowable heat flux. The ASME code prescribes a set of curves of allowable strain ε_T versus number of cycles N_d for failure due to creep–fatigue interaction. These curves (which are given in Figure 10.16 for 316 stainless steel) may be used to establish $\varepsilon_{th}^{max}(N_d) = \varepsilon_T(N_d)$ and then, from the above expression for thermal strain,

$$(q_p''\delta)^{max} = \frac{2\kappa(1-\nu)}{\alpha}\varepsilon_{th}^{max}(N_d)$$

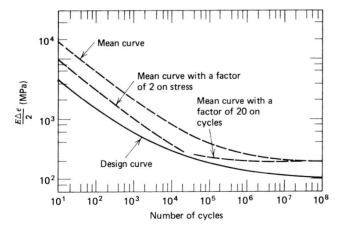

Figure 10.16 Fatigue design curve for type 316 stainless steel (Reproduced with permission from W.M. Stacey *et al.*, "US INTOR", Ga. Inst. Techn. report, 1981.)

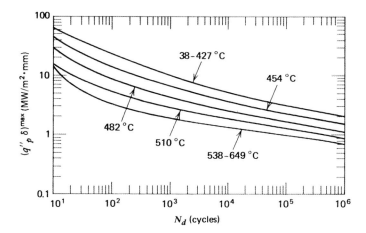

Figure 10.17 Maximum value of product of surface heat flux and wall thickness as a function of number of allowable cycles for 316 stainless steel. (Reproduced with permission from D.L. Smith *et al.*, Argonne National Lab. report, ANL/FPP-79-I, 1979.)

This quantity is plotted in Figure 10.17 for stainless steel. The maximum allowable surface heat flux is then found by using the above expression for $\delta_{\min}$,

$$(q_p'')^{\max} = \frac{(q_p''\delta)^{\max}}{\delta_{\min}} = \frac{2\kappa(1-\nu)}{\alpha\delta_{\min}}\varepsilon_{\mathrm{th}}^{\max}(N_d)$$

An upper limit for the allowable surface heat flux can be found by neglecting erosion in the evaluation of $\delta_{\min}$. Figure 10.18 illustrates this upper limit for a typical

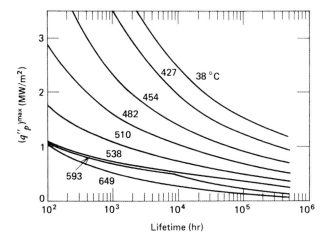

Figure 10.18 Maximum surface heat flux as a function of design lifetime for 316 stainless steel, $pR = 0.25$ MPa-m; 30-min burn pulse. (Reproduced with permission from D.L. Smith *et al.*, Argonne National Lab. report, ANL/FPP-79-I, 1979.)

value of $pR = 0.25$ (e.g., $Li - p \simeq 1$ MPa, $R = 0.25$ m; $H_2O - p = 10$ MPa, $R = 25$ mm) and for a cycle operating time of 0.5 hr. When erosion due to sputtering and disruptions is taken into account, δ_{min} can be considerably greater and the maximum allowable surface heat flux can be considerably lower.

Other materials, with better thermophysical properties, have higher allowable surface heat fluxes than stainless steel. Tungsten, tantalum, and alloys of copper and vanadium are of particular interest in this regard. Such materials almost certainly will be required for divertor collector plates and limiters.

A pulsed mode of operation leads to another potentially limiting phenomenon for surface components. Initial flaws, which are too small to be detected, will grow under a cyclic thermal stress loading until fracture of the tube wall occurs. The rate of crack growth is a function of the range of the stress intensity factor ΔK (MPa m$^{1/2}$) and temperature. The range of variation in the stress intensity factor over an operating cycle is

$$\Delta K \simeq 1.2 \Delta \sigma_b \sqrt{\pi a_0}$$

where σ_b is the average bending stress that would be present in the absence of the flaw, $\Delta \sigma_b$ is the additional bending stress associated with the cyclic loading, and a_0 is the initial radius of the flaw. For the surface components subjected to a pulsed surface heat flux q''_p, $\sigma_b = \sigma_{th}$ and $\Delta \sigma_b = \Delta \sigma_{th}$ depends upon difference in the temperature gradient over the operating cycle.

The rate of flaw growth can be represented by an equation of the form

$$\frac{da}{dN} (\text{mm/cycle}) = f_1(T)(\Delta K)^\gamma$$

where γ is of order unity. The crack growth rate for cold-worked 316 stainless steel is given in Figure 10.19.

10.5
ITER Blanket Test Modules

Testing of tritium breeding blanket modules (TBBMs) for the following DEMO reactors is one of ITER's primary objectives. During the D-T phase of operation, the reference operating mode of ITER will provide an average first-wall surface heat flux up to 0.27 MW/m^2 (0.5 MW/m^2 over 10% of first-wall), a 14 MeV neutron flux up to 0.78 MW/m^2, and a pulse length of 400 s with a 22% duty cycle. Pulse lengths up to 3000s, with a duty cycle up to 25%, could be achieved in the planned non-inductive operational scenario. The 400s reference pulse is longer than the thermal time constants of the TBBM subcomponents, so that relevant temperatures will be achieved. However, most tests relating to tritium release performance need >1000s pulses which will only be available in the non-inductive operation phase.

Three equatorial ports, each 1.75 m wide by 2.2 m high are allocated for TBBM test ports. These ports extend from the plasma-facing first wall for 20 cm into the surrounding shield.

The different ITER Parties (EU, Japan, Russia, USA, China, India, Korea) have different concepts for the tritium breeding blanket for their national DEMO reactor to

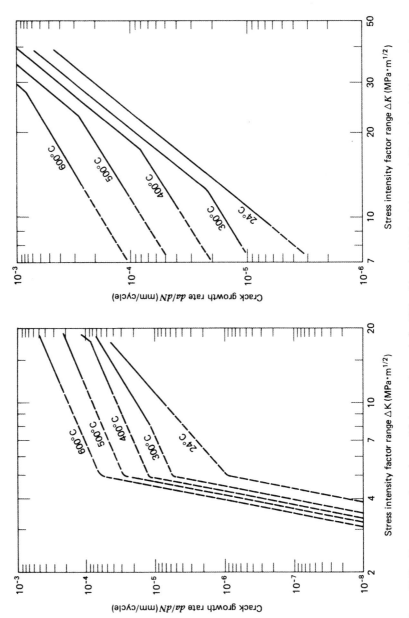

Figure 10.19 Temperature dependence of fatigue crack growth for 20% cold-worked 316 stainless steel. (Reproduced with permission from DOE/TIC-10 122, Dept. Energy report, 1980.)

follow ITER. However, there are some common requirements: tritium breeding self-sufficiency, first-wall neutron flux $>2\,MW/m^2$, first-wall heat flux 0.5–$1.0\,MW/m^2$, low-activation materials, thermal energy conversion efficiency $>30\%$.

All the ITER Parties are interested in helium-cooled ceramic tritium breeding blankets, with ferritic/martensitic steel structure and beryllium neutron multiplying material. The preferred ceramic breeder is either Li_2TiO_3 or Li_4SiO_4 in pebble-bed (diameter $\approx 1\,mm$) form. The beryllium neutron multiplier is also in pebble-bed form except for the Russian porous Be block design. A water-cooled version of the same concept is also proposed by Japan.

Liquid lithium-lead tritium breeding blankets with ferritic/martensitic steel are of interest to several of the Parties. The EU has selected a helium cooled version, while the USA and China have chosen dual-coolant versions with He cooling of the first-wall and structure but with lithium-lead self-cooling of the breeder zone.

Liquid lithium blanket concepts have been proposed by Russia and Korea. The Russian version is self-cooled with lithium and has vanadium alloy structure and beryllium neutron multiplier. The Korean version is He-cooled with ferritic/martensitic steel structure and a graphite pebble-bed reflector at the rear.

The ambitious Japanese DEMO program also includes more advanced designs requiring further R&D. These include a dual-cooled molten salt tritium breeding blanket, a water-cooled ceramic tritium breeding blanket, and a helium-cooled concept with SiC/SiC structure capable of operating at high temperatures to achieve high thermal conversion efficiency.

10.6
Fissile Production

A fusion reactor operating on the D-T cycle will produce a relatively large number of neutrons per unit of energy (17.6 MeV of fusion energy, leading to 20–25 MeV of thermal energy, per fusion neutron). With a modest amount of neutron multiplication via $(n, 2n)$, $(n, 3n)$, and so on reactions, a fusion reactor should be capable of breeding enough tritium for self-sufficiency and have excess neutrons which can be used for other purposes.

The fusion-fission hybrid concept would use these excess neutrons to produce fissile fuel via nuclear transmutation following neutron capture. The blanket surrounding the plasma chamber would contain a "fertile" material, uranium-238 or thorium-232. Both of these materials have a significant fast neutron capture cross section, which converts (via radioactive decay) the fertile isotopes into the isotopes plutonium-239 and uranium-233, respectively, which have large slow neutron fission cross sections and smaller but significant fast neutron fission cross sections. The fast neutron fission cross sections for the uranium and plutonium isotopes are given in Figures 10.20 and 10.21. (Refer to Figure 9.4 for a representative blanket neutron spectrum.)

The hybrid blanket would consist initially of natural or depleted uranium. The fissile material that was produced could be removed for fueling fission reactors or could be partially "burned" (fissioned) *in situ* in the hybrid blanket.

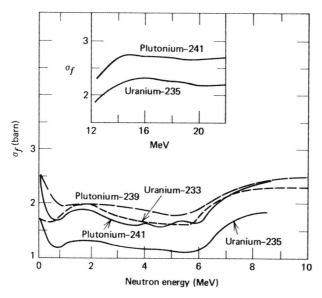

Figure 10.20 Fast fission cross section for fissile isotopes. (Reproduced with permission from L.J. Templin, ed., Argonne National Lab report ANL-5800, 1963.)

About 200 MeV and 2–3 neutrons are released in a fission event. Thus, the presence of fertile material (uranium-238 or thorium-232) in a fusion reactor blanket leads to a significant energy enhancement and neutron multiplication via the uranium-238 or thorium-232 (n, fission), (n, $2n$) and (n, $3n$) reactions. The energy enhancement is much greater with uranium-238 than with thorium-232 because of the larger fast neutron fission cross section of the former. If the fissile isotopes are left in the hybrid blanket, their fission contributes an additional energy enhancement and neutron multiplication. The larger fast neutron fission cross section for plutonium-239 relative to that of uranium-233 makes the energy enhancement greater for the former. Quite large values of the energy enhancement factor

$$M \equiv \frac{\text{total thermal energy deposited in blanket}}{\text{fusion neutron energy}}$$

are possible. $M \simeq 5$–10 can readily be achieved in blanket designs.

From an overall energy resource viewpoint it may be preferable to achieve a high fissile material production rate rather than a high energy multiplication in a hybrid blanket. In this case, suppression of the uranium-233 or plutonium-239 fission rate in the hybrid blanket must be achieved by removal from the blanket before a significant buildup of fissile material occurs. Mobile fuels such as slurries, molten salts, and balls have been considered.

Representative performance characteristics predicted for hybrid reactors are given in Table 10.11, and representative fueling characteristics of 1000-MW$_{el}$ fission reactors are given in Table 10.12. Based on these figures it appears that one hybrid

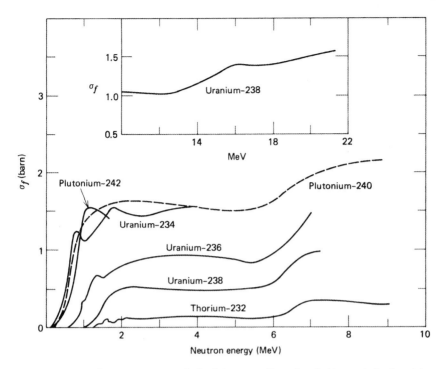

Figure 10.21 Fast fission cross section for fertile isotopes. (Reproduced with permission from L.J. Templin, ed., Argonne National Lab report ANL-5800, 1963.)

Table 10.11 Parameters for tritium-self-sufficient hybrids with low blanket energy Multiplication[a].

Parameter	Thorium/uranium	Uranium/plutonium
Blanket energy multiplication	1.5	5.0
Fissile-breeding ratio	0.5	1.5
Tritium-breeding ratio	1.0	1.0
Tritium inventory (kg)	5.0	5.0
Fusion power (MW)	1000	1000
Thermal power (MW$_{th}$)	1375	4000
Gross electricity output (MW$_{el}$)	550	1600
Fissile production rate (kg/yr)[b]	1630	4995
Required tritium fuel (kg/yr)[b]	0	0
Initial fissile loading (ton)	0	0

Reproduced with permission from S.I. Abdel-Khalik, *Nucl. Tech./Fusion*, **3**, 1983.

a) The number of breeding captures per fusion event, that is, the sum of fissile- and tritium-breeding ratios, given in this table represent upper bounds on hybrid performance parameters reported in the literature.

b) Figures arc based on 75% capacity factor.

Table 10.12 Representative fueling characteristics of 1000 MW$_{th}$ fission reactors[a].

Reactor type	LWR[b]	LWR	FBR[c]
Fissile fuel	Uranium-233	Plutonium	Plutonium
Conversion ratio	0.78	0.71	1.32
Initial core loading (kg)	1580	2150	3160
Discharge burnup (MWd/kg)	35	33	
Annual reload (kg)	720	1000	1480
Annual discharge (kg)	435	650	1690
Net annual makeup (kg)	285	350	—
Net annual excess (kg)	—	—	210

Reproduced with permission from S.I. Abdel-Khalik, *Nucl. Tech./Fusion*, **3**, 1983.
a)   From Working Group V, "International Nuclear Fuel Cycle Evaluation, Final Report," IAEA,
     Vienna, 1980. Assumed 75% capacity factor.
b)   Light-water reactor.
c)   Fast-breeder reactor.

reactor producing 1000 MW fusion power could support 5.7 or 3.7 1000 MW$_{th}$ light-water fission reactors (LWRs), depending upon whether or not the discharged fuel from the LWR was reprocessed to recover the fissile content, on the thorium/uranium fuel cycle. On the uranium/plutonium fuel cycle one hybrid reactor producing 1000 MW fusion power could support 14.3 or 5.0 1000 MW$_{th}$ LWRs. These support ratio estimates are probably too optimistic because of the approximations made in the hybrid fissile production rate calculations. However, these estimates are sufficiently accurate to establish the great potential that exists for using fusion reactors as fuel factories to supply LWRs. By contrast, several fast breeder reactors would be required to supply a single LWR.

10.7
Transmutation of Nuclear Waste

The spent fuel discharged from nuclear reactors in the USA is accumulating at a rate that would require a new Yucca Mountain type of long term (10^6 yrs) high level waste repository (HLWR) to be built every 30 years. The long-term requirement is imposed by the presence of transuranic elements which have been produced by neutron transmutation from ^{238}U (plutonium, americium, etc.), which have very long half-lives ($\sim 10^6$ yrs). All of these isotopes are fissionable in a fast neutron spectrum (i.e., a fast reactor) and could be almost completely destroyed if they were irradiated long enough in a nuclear reactor.

   The problem is that as the fissionable material is converted to fission products, which are parasitic neutron absorbers, it becomes impossible to maintain the neutron balance that sustains the critical neutron fission chain reaction. This can be offset up to point, limited by safety considerations, by removing other neutron absorbing material from the reactor as the fuel depletes. However, it is very difficult to maintain criticality as the fuel fissions beyond a small percentage. This problem could be solved by adding a variable strength neutron source which could be increased to maintain a constant fission power level as the fission fuel was destroyed, allowing much higher transuranic fuel burnup to be obtained.

Conceptual studies of a sub-critical transmutation reactor driven by a tokamak fusion neutron source based on the physics and technology database used for the ITER design have been performed. While it is theoretically possible to operate such a transmutation reactor long enough that >90% of the fissionable transuranic material was fissioned, it is unlikely that structural materials could be developed which could survive in such an extreme fast neutron fluence.

A commonly used estimate of the upper limit on radiation damage to advanced structural materials before failure is 200 dpa (which has been achieved with ferritic/ martensitic steels), which corresponds in the these designs to fissioning 24% of the original transuranic fuel before it had to be removed and reprocessed to remove the parasitic fission products and add fresh fuel. By reprocessing and recycling the fuel several times, greater than 90% burnup of the original fissionable material could be achieved. The resulting material sent to HLWRs would only produce 1–10% of the integral decay heat that would have been produced if the spent fuel originally discharged from a conventional nuclear reactor had been sent directly to the repository. Thus, there is a possibility of reducing HLWR requirements by a 10–100 fold by coupling fast transmutation reactors to variable-strength fusion neutron sources.

The tokamak fusion neutron source (described in Chapter 12) required for such a "closing the nuclear fuel cycle" mission has parameters that are comparable to or less than those of ITER ($Q_p = 2-5$, $\beta_N = 2.0-2.5$, $P_{fus} \approx 400$ MW), except for the practical requirement of higher availability $\geq 50\%$.

Problems

10.1. Consider a one-dimensional slab-type blanket consisting of 2-cm thick breeder elements cooled on both sides by water at $T_f = 300\,^{\circ}$C. The volumetric nuclear heating rate in the blanket is 5 W/cm^3 and the thermal conductivity is 5 W/m · K. The water is flowing at 5 m/sec. Calculate the temperature distribution across the slab breeder element. Take $D = 2.5$ cm for the equivalent cooling channel diameter.

10.2. Calculate the pressure drop in Problem 10.1 if the coolant channel is 2 m in length.

10.3. A breeding blanket is designed so that there is 0.25 m^2 of breeder, with nuclear heating density $q_b''' = 5$ W/cm^3, per coolant channel. Calculate the increase in the coolant (H$_2$O) temperature between inlet and exit if the channel is 2 m long and 5 cm in diameter and the water is flowing at 2.5 m/ sec. (Assume a constant density of $\rho = 900$ kg/m^3 and a heat capacity of $C_p = 4100$ J/kg K.)

10.4. Calculate the thermal and hoop stresses for the coolant tube in Problem 10.3 if it is made of stainless steel and is 1 cm thick, and if the water is at 14 MPa pressure.

10.5. Calculate the MHD pressure drop for liquid lithium flowing at 1 m/sec down a stainless steel tube of thickness 0.2 cm, diameter 2 cm, and length 2 m in a transverse magnetic field $B_\perp = 5$ T.

10.6. Calculate the thermal strain and stress in a stainless steel and in a vanadium coolant tube of radius 2 cm and thickness 0.25 cm carrying water coolant at 15 MPa if the heat flux incident on the tube from the surrounding blanket region is 2 MW/m².

10.7. Consider a tokamak plasma with major radius $R = 6.5$ m and minor radius $a = 2.0$ m that produces 2500 MW$_{th}$ fusion power. The plasma chamber wall has a radius $a_w = 2.2$ m and is equipped with a limiter which has an area 10% that of the wall. Calculate the surface heat flux to the wall and limiter as a function of the charged particle fraction of the surface heat flux $f_\pm$, for $0.1 \leq f_\pm \leq 0.9$.

10.8. Calculate the total volumetric neutron heat deposition in the blanket for Problem 10.7. (Assume 20 MeV per fusion neutron is deposited in the blanket.) Calculate the required mass flow rale $\dot{W}$ (kg/sec) of H₂O coolant at 10 MPa if $T_{in} = 100\,°C$ and $T_{out} = 300\,°C$.

10.9. Assume that the average charge-exchange D-T flux to the first wall of the tokamak in Problem 10.7 is $3.5 \times 10^{19}\,m^{-2}\,sec^{-1}$ and that the average particle energy is 100 eV. Assume that the first wall consists of stainless-steel coolant tubes. Calculate the thickness that would be eroded away in 2 yr of operation. If the tubes have radius $R = 2.5$ cm, carry H₂O at 10 MPa and operate at 400 °C, calculate the minimum initial thickness that would be required to achieve 5 yr of operation. Assume that the ASME code prescribes $S_{mt} = \frac{2}{3} S_y$ ($S_y =$ yield stress) at operating temperature.

10.10. Calculate the thermal stress in the first-wall coolant tube of Problems 10.7–10.9 at initial operation and after 2 yr, assuming that the charged particle fraction of the heat flus is $f_\pm = 0.8$.

10.11. Calculate the thermal strain after 1.5 yr of operation for Problem 10.10. Use this to estimate the number of cycles to fatigue failure N_d. What is the minimum operating cycle time that allows 5 yr operation before fatigue failure?

10.12. Calculate the size of a crack with initial size $a_0 = 0.5$ cm for the first wall of Problems 10.7–10.11 after 2 yr operation for an operating cycle of 1 per hour. Assume that the first-wall coolant tube temperature gradient vanishes between cycles.

11
Tritium and Vacuum

Tritium, an isotope of hydrogen with atomic mass equal to 3, is a radioactive element that decays by β emission with a half-life of 12.3 yr. Even if available in sufficient quantity, tritium would be an expensive fuel. Like the lighter isotopes of hydrogen (protium and deuterium), tritium can diffuse rapidly through most materials at elevated temperatures, and hence its containment becomes a major issue of concern to the design of fusion reactor fuel handling and recovery systems. For virtually all D-T fusion reactor concepts, there is a requirement to pass relatively large quantities of tritium through the vacuum pumping and fuel handling systems. Thus, by virtue of their direct contact with the tritium-containing fuel stream, these systems must be designed to serve as the primary containment boundary.

11.1
Vacuum System

Fusion systems have substantial vacuum pumping requirements. The plasma chamber, which will typically be several hundred cubic meters in volume, must be evacuated down to $\sim 10^{-7}$ torr (1 Pa = 0.008 torr, 1 atm = 760 torr) at the initiation of an operating period, and pumped down from $\sim 10^{-3}$ torr to $\sim 10^{-5}$ torr in a relatively short time between operating cycles. The pressure in the ducts and cavities in the neutral beam and RF heating systems, with volumes of several tens of cubic meters, must be maintained at $\sim 10^{-7}$ torr.

11.1.1
Vacuum Pumping

The factors involved in pumping a chamber down to achieve a high vacuum (i.e., low pressure) may be understood by considering a chamber with volume V_c, pressure p_c, and gas source Q_c connected via a pumping duct of length L to a pump with ultimate obtainable pressure p_u (limited by backflow) and theoretical pumping speed S_t, as depicted in Figure 11.1.

Fusion: Second, Completely Revised and Enlarged edition. Weston M. Stacey
Copyright © 2010 Wiley-VCH Verlag GmbH & Co. KGaA, Weinheim
ISBN: 978-3-527-40967-9

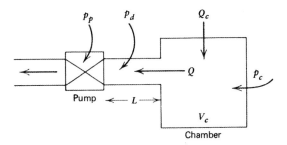

Figure 11.1 Chamber and pump.

The rate Q at which gas is pumped out of the chamber is related to the theoretical pumping speed of the pump, S_t, and the difference between the actual p_p and ultimate obtainable p_u pressures at the pump by the equation

$$Q = S_t(p_p - p_u)$$

This "throughput" rate can also be expressed in terms of the difference between the pressures in the chamber p_c and at the pump p_p and the conductance C of the duct,

$$Q = C(p_c - p_p)$$

The rate at which the chamber is evacuated is

$$V_c \frac{dp_c}{dt} = Q_c - Q = Q_c - \frac{S_t C}{S_t + C}(p_c - p_u)$$

which, when C is constant, has the solution

$$p_c(t) = p_c(0)e^{-t/\tau_c} + \left(p_u + \frac{Q_c}{V_c}\tau_c\right)(1 - e^{-t/\tau_c})$$

The characteristic time

$$\tau_c \equiv \frac{(S_t + C)V_c}{S_t C}$$

determines the evacuation rate. Thus, the chamber pressure decreases exponentially to an asymptotic value,

$$p(\infty) = p_u + \frac{Q_c}{V_c}\tau_c$$

The conductance C must be evaluated from gas dynamics. It depends upon whether molecule – wall interactions are dominant (the molecular flow regime) or molecule – molecule interactions are dominant (the viscous flow regime). Defining D as the diameter or other characteristic cross-sectional dimension of the duct and λ as the molecular mean free path between molecule – molecule interactions, the flow regimes are defined by

Table 11.1 Gas constant and viscosity (1 atm, 273 K) for several gases.

Gas	k_1 (Pa m)	v_0 (10^{-7} Pa sec)
Air	0.0068	170.8
H_2	0.0141	83.5
He	0.0195	186.0
N_2	0.0081	166.6
O_2	0.0092	189.0
Ar	0.0093	209.6
H_2O (vapor)	0.0127	

Reproduced by permission from T.J. Dolan, *Fusion Research*, Vol. III, Pergamon Press, 1982.

$$\lambda/D \gtrsim 1 \qquad \text{molecular flow}$$
$$\lambda/D \lesssim 0.01 \qquad \text{viscous flow}$$

The molecular mean free path may be expressed as

$$\lambda = \frac{k_1}{p}$$

where p (Pa) is the gas pressure and k_1 is the gas constant, values of which are given in Table 11.1 for several gases. The viscosity v_0 at atmospheric pressure and 273 K is also given.

For the viscous flow regime,

$$C(\text{m}^3/\text{sec}) = \begin{cases} 1430 \dfrac{D^4 p_d}{L} \dfrac{v_{\text{air}}}{v} & \text{circular duct} \\[2ex] 2070 \dfrac{Y p_d a^2 b^2}{L} \dfrac{v_{\text{air}}}{v} & \text{rectangular duct, with dimensions } a \text{ and } b \end{cases}$$

where $p_d = \frac{1}{2}(p_p + p_c)$, Y is a parameter that depends upon the duct dimensions, as follows:

a/b	1.0	0.6	0.4	0.2	0.1
Y	1.0	0.9	0.71	0.42	0.23

v is the viscosity. The viscosity at other temperatures scales from the values in Table 11.1 as

$$v(T) = v_0 \left(\frac{T}{273}\right)^{1/2}$$

For the molecular flow regime,

$$C(\text{m}^3/\text{sec}) = \begin{cases} 122 \dfrac{D^3}{L(1 + 4D/3L)} & \text{circular duct} \\[2ex] 309 \dfrac{Ka^2 b^2}{L(a+b)} & \text{rectangular duct } a \times b \end{cases}$$

where K depends upon the duct dimensions as follows:

a/b	1.0	0.5	0.33	0.2	0.1
K	1.11	1.15	1.20	1.30	1.44

For a multistaged pumping configuration where several pumps are connected in series by ducts of different conductances, an effective conductance can be defined by

$$\frac{1}{C_{\text{eff}}} = \frac{1}{C_1} + \frac{1}{C_2} + \cdots$$

When the chamber is evacuated by pumping through several parallel ducts, an effective conductance is defined by

$$C_{\text{eff}} = C_1 + C_2 + \cdots$$

Several types of pumps are used in fusion experiments and/or are under consideration for use in future reactors. Cryosorption and cryocondensation pumps or panels are expected to have widespread application in future fusion devices. In a cryosorption pump, sievelike pellets are cooled down to as low as 77 K by liquid nitrogen or as low as 4 K by liquid helium. Gases flowing over these pellets are strongly adsorbed. Cryocondensation pumps consist of metal plates cooled by liquid nitrogen or helium. Gas flowing over these plates will condense if the plate temperature is lower than the gas condensation temperature. Both types of cryopumps can be regenerated by isolating them from the chamber and warming them up, which drives off the adsorbed or condensed gases. Liquid helium cooled cryocondensation pumps will pump H, D, T, and other gases found in a fusion system, except helium. Liquid helium coolant must be used in cryosorption pumps to achieve helium gas pumping. Both types of cryopumps must be protected from nuclear and plasma radiation heating in order to operate effectively.

Getter, or sublimation, pumps consist of plates of chemically active metals such as titanium or zirconium–aluminum that chemically react with gases (e.g., hydride formation with H, D, T). These pumps can operate at elevated temperatures, but they cannot pump helium or other inert gases. Once the surface is covered with reacted molecules, the pump must be regenerated by removing the reacted surface film or by depositing a new metallic surface film. Both processes are relatively time consuming and lead to the accumulation of an additional tritium inventory.

Turbomolecular pumps function like a fan. Disks rotating at a high velocity impart momentum directed away from the chamber to gas molecules.

The principle of a diffusion pump is to spray a fluid (oil or mercury) through a nozzle to impart momentum away from the chamber to molecules of the gas to be pumped. Oil would contaminate the gas being pumped; thus only mercury diffusion pumps could be used in fusion reactors. Even with mercury, cold traps and baffles must be placed between the pump and the chamber to prevent contamination.

A variety of mechanical pumps are available. Rotary mechanical pumps consist of a rotating piston with vanes which sweep the gas from an inlet port to an exit port. A Roots-type mechanical pump consists of two meshing gears which sweep the gas

Table 11.2 Vacuum pump characteristics.

Pump type	Maximum pump speed	Minimum pressure	
		(Pa)	(torr)
Cryocondensation	$10\ell/\sec\,cm^2$	10^{-8}	8×10^{-11}
Cryosorption	$3\ell/\sec\,cm^2$	10^{-1}	8×10^{-4}
Getters (Zr-Al)	$5\ell/\sec\,cm^2$	10^{-7}	8×10^{-10}
Turbomolecular	$2 \times 10^4\,l/\sec$	10^{-6}	8×10^{-9}
Mercury diffusion	Small	10^{-6}	8×10^{-9}
Rotary blower	$10^3\ell/\sec$	10^{-1}	8×10^{-4}
Mechanical (oil-free)	$10^2\ell/\sec$	10^{-2}	8×10^{-5}

Reproduced by permission from "INTOR-Zero Phase," IAEA report STI/PUB/556, 1980.

from an inlet port to an exit port as they rotate. Mechanical pumps must be used in tandem with other pumps such as "appendage," or "roughing," pumps to reduce the pressure before the mechanical pumps take over. Pumps with oil lubricants and sealants generally should be avoided in a tritium environment.

Characteristics of the various types of pumps are given in Table 11.2.

The minimum obtainable pressure depends upon the gas source into the chamber as well as upon the ultimate pumping speed. The gas source into the chamber consists of (i) outgassing of residual gases in the materials of the chamber wall, (ii) desorption of gases adsorbed on the surface; and (iii) gases leaking in through the primary vacuum containment. Many of the gases originally present in the chamber materials may be removed by heating the walls or by subjecting the walls to low-temperature plasma discharges. For example, the outgassing rate of untreated stainless steel or Inconel is in the range 10^{-4}–10^{-7} Pa·m^3/sec·m^2. These rates can be reduced to less than 10^{-8} Pa·m^3/sec·m^2 by "baking" the material at 500 °C for 100–200 hr.

11.1.2
Vacuum Chamber Topology

There are many options for the topology of the primary vacuum containment for the plasma chamber. (The primary vacuum containment bears the pressure differential from atmospheric to plasma chamber pressures.) In present tokamak experiments the vacuum containment is a vessel immediately surrounding the plasma. Such a location is impractical for a fusion reactor because of neutron damage and thermal cycling (see Section 10.4).

Several options that have been considered for primary vacuum containment in a tokamak reactor are illustrated in Figure 11.2. In Figure 11.2a the torus and the superconducting coils are contained within a common vacuum vessel. In Figure 11.2b a vacuum dome encloses the entire reactor. In Figure 11.2c the primary vacuum containment for the plasma is located just behind the blanket; a separate

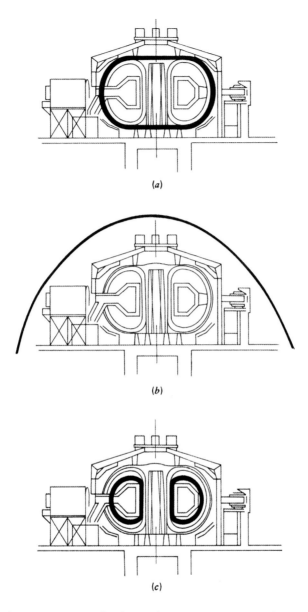

(a)

(b)

(c)

Figure 11.2 Vacuum chamber topology options. (Reproduced by permission from W. M. Stacey *et al.*, "USINTOR", Ga. Inst. Techn. report, 1980.)

vacuum containment is required for the superconducting magnets in this case. Options *a* and *b* would have a secondary vacuum containment at the plasma chamber which achieved the high vacuum of the plasma chamber, but the atmospheric pressure would be taken at the primary containment. Option *b* has as disadvantages the necessity of breaking the primary containment to gain access to the externals of the reactor and the absence of any control of tritium which may diffuse out of the blanket, short of the primary containment. Option *a* leads to relatively complex engineering designs. Thus, at present some variant of option *c* is favored for tokamak reactors.

11.2
Tritium Fuel Cycle Processing[1]

The tritium fuel cycle system must recover the tritium from the exhaust gas that is pumped from the plasma and from the breeding blanket, and process this tritium so that it can be introduced into the plasma chamber to fuel the fusion reactor.

The fuel cycle system can be divided into a fuel purification system and an isotope separation system. The fuel purification system separates all the gaseous chemical impurities, including helium, from the mixture of hydrogen isotopes which has been pumped out of the plasma chamber or extracted from the blanket. It also recovers all the tritium which is chemically combined with these impurities. The isotope separation system separates the purified fuel mixture of hydrogen isotopes into individual streams with different concentrations of D and T isotopes (as required for fuel preparation and injection), separates the unwanted H isotope (mainly arising from in-leakage) from the fuel mixture so that it can be discharged as a tritium-free waste stream, and separates the unwanted H and T isotopes from the isotopically pure D_2 used in NBI circuits.

The exhaust gas from the plasma chamber is expected to have roughly the following atomic composition:

D, T	$\sim$85–90%
H	$\sim$1–10%
^{4}He	$\sim$5–10%
O, N, and C	$\sim$1%
Metallic impurities	$\sim$0.01%

The metallic impurities can be removed from the exhaust by electrostatic precipitation, thus leaving a stream of hydrogen atoms and gaseous impurity atoms (C, O, N, He, etc.). Oxygen, nitrogen, and carbon are present mainly in the form of the chemical compounds water [DTO], ammonia [N(D, T)$_3$], and methane [CD_2T_2], respectively. After separation from the main stream, all these compounds must be chemically dissociated to separate and recover all the chemically combined tritium and discharge the impurity atoms to a tritium-free waste.

---

1) Portions of this section were extracted directly from the IAEA report STI/PUB/556, with permission of the IAEA.

Various methods are known to effect the separation of the remaining impurities from hydrogen. One of the well-known ones is the use of palladium–silver alloy membranes, but with this method all the chemically combined tritium and also some molecular tritium pass out with the waste gas stream. Thus, although the main fuel stream is purified and the impurities are concentrated, the more difficult separations still remain to be performed in order to detritiate the waste. In this kind of unit, the flow of gas to be purified is sent into a reactor equipped with thin membranes. The mechanism inside the reactor is the following. Molecules of hydrogen at the upstream surface of the membrane dissociate into atoms. These atoms migrate through the membrane by bulk diffusion. At the other surface the hydrogen atoms recombine to form molecules. The whole mechanism is a physicochemical phenomenon which, in principle, should yield a stream of pure hydrogen isotopes. On the other hand, in order to have a high enough flow rate, the membranes must be heated to temperatures of around 450–500 °C. The combination of this relatively high temperature with a hydrogen pressure higher than 1 atm above the membranes may cause a large tritium permeation through the walls of the membrane housing. Also, the membrane can be "poisoned" by prolonged exposure to some impurities.

Another method used to purify and recover the chemically combined tritium is by chemical reaction with hot metals (e.g., uranium or titanium) at temperatures high enough for the gaseous impurities to form solid oxides, nitrides, and carbides and for solid hydrides not to be formed so that all the hydrogen isotopes remain in the gas phase. But one must avoid any accidental introduction of oxygen into such beds in order to prevent a fire. The helium and other inert gases are not removed by hot metals and will need to be separated by physical processes. This hot metal technology involves the problems of diffusion of tritium through metal walls at high temperatures and the disposal of tritium contaminated solid wastes.

A third solution to the purification and tritium recovery problem, which requires only gaseous wastes with very low tritium content, is based mainly on cryogenic gas separation techniques and on electrolysis of DTO. It is convenient to combine this with cryogenic distillation for isotope separation and cryoseparation of the helium. The chemically combined tritium is oxidized to convert all the tritium into water (DTO), which is then cryoseparated and electrolyzed to recover all the tritium. The impurity atoms are then all in the form of CO_2, O_2, and N_2, which are nontoxic and, after cryoseparation from DTO, can be discharged as gaseous waste. There is no need for containment of the main fuel stream at high temperatures, and tritium partial pressures are low for the oxidation step.

The "hydrogen" stream which emerges from the above purification processes consists, in general, of a mixture of six molecular species: H_2, HD, HT, D_2, DT, and T_2. For a given atom fraction of H, D, and T the molecular composition will depend on whether the molecules are in thermodynamic equilibrium. This equilibrium (isotopic) can be attained with a catalyst at about 20–200 °C. If it is assumed that the D-T fuel will be in high-temperature molecular equilibrium (isotopic), then a 50/50 D-T mixture containing a small amount of hydrogen will have a molecular composition of

DT ≅ 50%

D_2 ≅ 25%

T_2 ≅ 25%

D = HT, small

H_2 very small

Thus, hydrogen (cryogenic) distillation, for certain separation duties, needs to be combined with chemical equilibration (20–200 °C) of the molecular composition. For example, (i) to promote D_2 + HT → HD + DT to split HT when it is necessary to extract more than half of the hydrogen from the fuel, and (ii) to promote 2DT → D_2 + T_2 should it be necessary to split more than half of the D-T fuel into D_2 and T_2 streams.

Hydrogen distillation is the process with which most experience with high tritium concentrations exists, and it is also favored because its high separation factor leads to a lower tritium inventory. Problems expected in hydrogen distillation are concerned mainly with controlling the operation of the several interlinked columns. Classical (isotope separation) theory predicts that, for a given separation duty and the known separation factors for hydrogen distillation and water distillation (at an optimum reduced pressure), the tritium inventory (liquid holdup) is about two orders of magnitude higher for water distillation, if we ignore the advantages of continuous molecular equilibration in water. There is also the need to develop electrolytic cells for water distillation (for converting all product streams back to hydrogen) which have a sufficiently low liquid inventory and are made of materials which do not suffer any radiation damage from the high tritium concentrations.

Another method of separating hydrogen isotopes is the countercurrent mass diffusion process. The separation is carried out in very simple columns at a pressure of 1 atm and a temperature of about 100 °C. The main construction of the elements of the column consists of a grid cylinder and a sprayer. A nonhydrogenous inert vapor is used as a carrier gas. This gas enters through the sprayer, crosses the grid cylinder, and is condensed onto the walls of a cylindrical condenser, which is cooled with water on the outside. The vapor continuously presses the gas to be separated to the surface of the condenser so that the gas diffuses into the counterflow of the vapor within the grid cylinder, wherein the primary separation takes place. With this method the primary separation factor is equal to 1.72 for the mixture H_2–D_2 and is equal to 1.34 for the mixture D_2–T_2. In the process of gas circulation along the column axis, a secondary countercurrent separation occurs. However, the secondary separation factor is equal to the primary separation factor at the column height, which is about equal to its diameter. The main advantage of this method lies in the fact that no preliminary purification is required. During the separation of the hydrogen isotopes, all the impurities are almost completely separated.

Bred tritium is expected to be removed from a solid breeder blanket by means of a helium purge stream. Taking into account the possibility of a mixture of T_2 and T_2O being present in the purge gas, the first step of a tritium-breeding processing system consists of a conversion of the T_2 and HT into water by passing it through a catalytic reactor in which, after injection of free oxygen, hydrogen is combined with oxygen over a DEOXO type catalyst at 450 K. Thereafter, helium has to be cooled to a temperature of 80 K over molecular sieve beds to remove all its water content. Following the separation of the sweep gas from the water vapor, helium is heated again to room temperature before being sent back into the breeding blanket. When required, the cold trap is heated to a temperature that makes it possible to recover its water content in liquid form. This water is transferred to an electrolytic cell where its decomposition into tritium and oxygen takes place. The recovered tritium is directed to the fuel clean-up system in the main stream, and some of the recovered oxygen is sent back to the catalytic bed.

With liquid breeding blankets, such as liquid lithium, the extraction of tritium could be achieved by a continuous method. The most promising methods that have been considered are: permeation of tritium through a "window" of highly permeable metal like niobium- or vanadium-base alloys, absorption of tritium by exothermic solid getters like yttrium and zirconium, extraction of tritium from liquid lithium into mixtures of liquid lithium halides (LiCl-LiF or LiF-LiCl-LiBr), and distillation.

Once the tritium is processed, it must be stored to await reuse. Gaseous storage could be implemented, but in this case the gas must be kept at a pressure lower than atmospheric pressure to prevent any leaks of tritium out of the container. The container itself must be stored in an adequately ventilated secondary container. Tritium storage as uranium or titanium hydride could also be used. This method involves heating the tritide to a temperature between 600 and 800 °C to recover the tritium, but at this high temperature we find again the problem of tritium permeation through the container walls.

11.3
Tritium Permeation

Hydrogen (H, D, T) diffuses readily in many metals. This poses the potential problem of tritium from the plasma or from the breeding blanket getting into the coolant and being transported out of the reactor. In the case of water coolant, recovery of the tritium from the water represents a considerable expense. An additional potential problem in this regard is the soak-in of tritium into structural components and the consequent buildup of an unavailable and potentially hazardous tritium inventory in the reactor structure. Finally, there is the possibility that tritium from the breeding blanket could diffuse through the vacuum boundary and escape to contaminate the reactor hall.

The diffusion of tritium in bulk material can be described by the particle balance equation

$$\frac{\partial J}{\partial x} + S = 0$$

where J is the current and S is the volumetric source, if any is present.

The tritium current J is related to the tritium concentration C gradient and the material temperature gradient by Fick's law,

$$J = -D\frac{\partial C}{\partial x} + \frac{CQ^*}{kT^2}\frac{\partial T}{\partial x}$$

where Q^* is the heat of transport, k is the Maxwell–Boltzmann constant, D is the diffusion coefficient

$$D \simeq D_0 e^{-E_D/kT}$$

and E_D is the activation energy for migration. For stainless steel $D_0 = 0.085$ cm^2/sec, $E_D = 0.61$ eV, and $Q^* = -0.065$ eV. The solution to the diffusion equation depends sensitively upon the source term S and the boundary conditions.

For the situation in which a tritium gas pressure p_1 builds up on one side of a metallic component (e.g., tube) with area A and thickness t, the source term is zero within the component. The permeation rate through such a component can be expressed as

$$\phi(\text{Ci}/\text{d}) = \frac{1}{\sqrt{3}}K\left(\frac{A}{t}\right)(p_1^{1/2} - p_2^{1/2})$$

where p_2 is the tritium pressure on the backside. The temperature-dependent function $K/\sqrt{3}$ is plotted in Figure 11.3 for several metals. This quantity can be reduced by one or more orders of magnitude if an oxide layer develops on the backside of the metal layer.

The components facing the plasma are bombarded with energetic tritium ions and atoms which penetrate the surface several micrometers and constitute a rather large "implantation source" in the diffusion equation. This source produces a large tritium concentration gradient near the surface on the plasma side, which leads to a high tritium flux to that surface. However, the tritium atoms arriving at this surface must combine into molecules before they can leave the surface and return to the plasma. The recombination coefficient is given by

$$k_r = \frac{5.25 \times 10^{25}}{(C_0^*)^2 \sqrt{T}} \alpha \exp\left(\frac{E_s - E_D}{kT}\right)$$

where E_s is the heat of solution, C_0^* is the Sievert's constant, and α is the molecular sticking coefficient, which is a function of the surface condition (composition, smoothness, etc.). For stainless steel $E_s = 0.091$ eV, $C_0 = 9 \times 10^{-4}$ atm$^{-1/2}$, and values of the sticking coefficient in the range $5 \times 10^{-5} \le \alpha \le 5 \times 10^{-1}$ have been observed. Neutron damage trapping of tritium also affects the permeation.

The effect of the implantation of energetic tritons is to enhance the permeation rate over that indicated in Figure 11.3. Calculations have been carried out for a representative stainless-steel first wall for an ETR with area 400 m^2 and thickness 0.5 cm subjected to a D-T particle flux $\phi = 2.5 \times 10^{16}$ cm^{-2} sec^{-1} of 200-eV average energy and cooled to 180 °C on the plasma side and 100 °C on the coolant side. The buildups

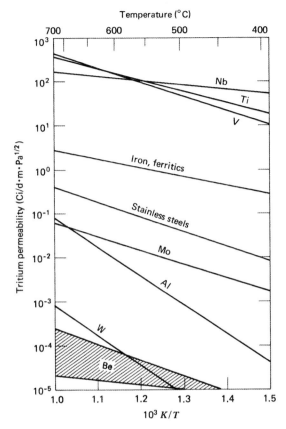

Figure 11.3 Tritium permeability of several materials as a function of temperature. (Reproduced by permission from W. M. Stacey *et al.*, "US INTOR", Ga. Inst. Techn. report, 1981.)

of the permeation rate and the tritium inventory in the wall are shown for continuous operation, in Figure 11.4 for several values of the molecular sticking coefficient α. The permeation rate would be considerably higher at the 400–500 °C temperatures of a fusion reactor first wall. Thus, tritium permeation through the first wall into the coolant is a potential problem area, and the development of permeation barriers seems necessary.

11.4
Tritium Retention in Plasma-Facing Components

Tritium retention in plasma-facing components is a major issue for future fusion devices (such as ITER) that will be fueled with deuterium-tritium mixtures. The buildup of tritium in plasma-facing components due to the implantation of incident

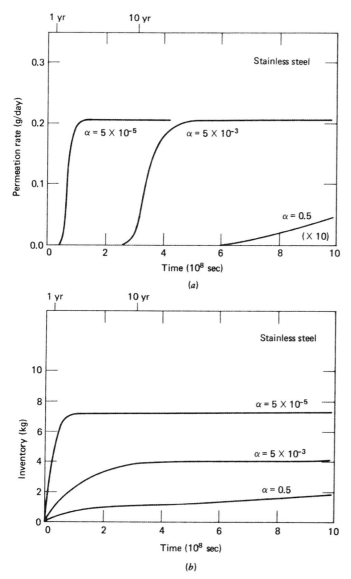

Figure 11.4 Effects of molecular sticking coefficient α on (a) permeation rate and (b) inventory in stainless steel. Wall area $= 400\,m^2$ and thickness $= 0.5\,cm$. (Note the multiplication factors for the various permeation rate curves.) (Reproduced with permission from W. M. Stacey *et al.*, US FED-INTOR", Ga. Inst. Techn. report, 1982.)

fluxes from the plasma is of course an obvious source of concern for the divertor collection targets and to a lesser extent for regions of the first-wall. Less obvious is the co-deposition of tritium implanted in material that has been sputtered or otherwise eroded from the first wall or divertor plate.

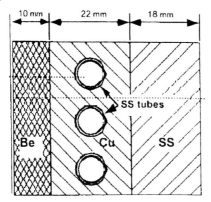

Figure 11.5 ITER first-wall design (Reproduced with permission from Physica Scripta – Figure 2 from the Federici, Brooks Phys. Scrip. T91, 76 (2001) paper.)

The ITER plasma-facing component configuration illustrates the range of tritium retention issues that are being addressed. The present design of the ITER first wall consists of a water-cooled copper layer bonded onto a stainless steel structure, with a beryllium coating facing the plasma, as depicted in Figure 11.5. The divertor target plate has a similar configuration, except that the coating is tungsten for the most part. Near the peak heat flux locations a carbon-fiber-composite (CFC) coating is used because it sublimes (rather than melts) during disruption thermal quenches and giant ELMs. The erosion and redeposition of this carbon, along with the implanted tritium, is a concern.

Intense co-deposition of deuterium with carbon is found in many present tokamaks, in regions which are shielded from ion fluxes but are near carbon surfaces with high incident deuterium ion fluxes. Since ions cannot impinge on these shielded surfaces, this carbon deposition can only be due to the diffusion of carbon atoms or molecules resulting from hydrocarbons released from carbon surfaces.

The inventories of hydrogenic species resulting from known incident ion fluxes have been measured for beryllium and tungsten (and other materials). With beryllium surfaces, the inventory saturates and is insensitive to continued incident ion flux. With tungsten, the inventory of implanted hydrogen not only saturates but actually decreases with continued incident ion fluxes (i.e. continued hydrogen ion influx stimulates reemission and leads to a lowered hydrogen inventory). Thus, the major concern for tritium retention in the plasma-facing wall material in ITER is the co-deposition of tritium with carbon that is eroded and redeposited. Current analysis indicates that the tritium co-deposition rate with the carbon eroded from the divertor target may be of the order of 2 grams for each 400 s, 500 MW pulse, while the tritium co-deposition rate with the beryllium that is eroded from the first-wall may be less than 0.5 gram per pulse.

Safety considerations suggest an allowable upper limit of 350 grams. The reference operating scenario assumes 12,000 of these pulses, so it will be necessary to remove the co-deposited tritium. Removal of the co-deposited layers could be accomplished by thermo-oxidative plasma erosion at temperatures $>250\,°C$ or by physical means, but the efficiency of this process is not yet established.

11.5
Tritium Breeding and Inventory

The tritium breeding rate in a fusion reactor blanket must be sufficient not only to replenish the tritium consumed by that plant, but also to provide for an excess to start up other plants. Allowance must be made for the radioactive decay of tritium and for the holdup of tritium in the blanket breeder and structural materials and the loss of tritium to the coolant via permeation through the first wall.

The rate equation governing the tritium inventory I_B (kg) in the blanket is

$$\dot{I}_B = \dot{N}_B - \left(\frac{1}{\tau_D} + \frac{1}{\tau_B} + \frac{1}{\tau_{LB}}\right) I_B \equiv \dot{N}_B - \frac{1}{\tau^*} I_B \qquad I_B(t=0) = 0$$

where $\dot{N}_B$ is the tritium production rate in the blanket due to neutron capture in lithium, $\tau_D = \lambda_D^{-1} = 18$ yr is the reciprocal of the radioactive decay time constant, τ_B is the mean residence time of bred tritium in the blanket, and τ_{LB} is the mean time for unrecoverable loss of tritium from the blanket into the coolant, structure, and so on. This equation has the solution

$$I_B(t) = \dot{N}_B \tau^* (1 - e^{-t/\tau^*})$$

The tritium inventory I_{ex} (kg), in the tritium processing and storage system external to the blanket is governed by the equation

$$\dot{I}_{ex} = \tau_B^{-1} I_B - \dot{N}_p (1+\varepsilon) - \left(\frac{1}{\tau_D} + \frac{1}{\tau_{Lex}}\right) I_{ex} \qquad I_{ex}(t=0) = I_0$$

where $\dot{N}_p$ is the rate of tritium consumption in the plasma, $\dot{N}_p \varepsilon$ is the rate of tritium loss by permeation through surface components into the coolant, and $\tau_{L\,ex}$ is the mean time against unrecoverable loss in the processing and storage system. Defining

$$\frac{1}{\tau^+} = \frac{1}{\tau_D} + \frac{1}{\tau_{Lex}}$$

the solution for the external tritium inventory is

$$I_{ex}(t) =$$
$$\frac{\tau^*}{\tau^+} \dot{N}_B \{\tau^+ (1 - e^{-t/\tau^+}) - \tau_B e^{-t/\tau^+} (1 - e^{-t/\tau_B})\} - \dot{N}_p \tau^+ (1 - e^{-t/\tau^+}) + I_0 e^{-t/\tau^+}$$

The conventional definition of the tritium breeding ratio TBR as the ratio of the tritium production rate to the tritium consumption rate by fusion is

$$\text{TBR} \equiv \frac{\dot{N}_B}{\dot{N}_p}$$

In order for the blanket to produce enough recoverable tritium to supply the tritium for that reactor,

$$\dot{N}_B \geq \dot{N}_p(1+\varepsilon) + \left(\frac{1}{\tau_D} + \frac{1}{\tau_{LB}}\right)I_B + \left(\frac{1}{\tau_D} + \frac{1}{\tau_{Lex}}\right)I_{ex}$$

or

$$\mathrm{TBR} \equiv \frac{\dot{N}_B}{\dot{N}_p} \geq 1+\varepsilon + \frac{1}{\dot{N}_p}\left[\left(\frac{1}{\tau_D} + \frac{1}{\tau_{LB}}\right)I_B + \left(\frac{1}{\tau_D} + \frac{1}{\tau_{Lex}}\right)I_{ex}\right]$$

A convenient way to characterize the capability to produce excess tritium to start up other fusion reactors is in terms of the doubling time, defined as the time required to double the external inventory, that is, by the relation

$$I_{ex}(t_2) = 2I_0$$

An optimistic lower limit on the doubling time associated with a given value of the TBR can be estimated by neglecting unrecoverable losses (namely, $\varepsilon \to 0, \tau_{LB} \to \infty$, $\tau_{L\,ex} \to \infty$). In this case the doubling time only depends upon the TBR and τ_B. A more convenient characterization is in terms of TBR and I_B, as shown in Figure 11.6. Accounting for unrecoverable losses increases the doubling time.

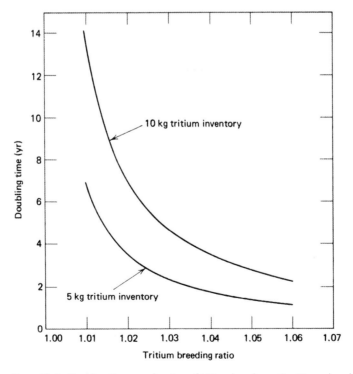

Figure 11.6 Doubling time as a function of tritium breeding ratio. (Reproduced with permission from C. C. Baker *et al.*, Argonne National Laboratory report, ANL/FPP-80-I, 1980.)

Problems

11.1. Consider a plasma chamber of volume 400 m³ in which there is a plasma with $n_D = n_t = 10^{20}$ m⁻³. Upon shutdown of the plasma burn, the D and the T cool to the chamber wall temperature of 400 °C and form D_2, DT, and T_2 molecules. Calculate the pressure of the exhaust gas in the chamber after the shutdown.

11.2. The plasma chamber in Problem 11.1 is pumped through 12 circular ducts, each of diameter 0.5 m, by cryocondensation panels with ultimate pressure 10^{-11} torr. Calculate the required pumping speed to evacuate the chamber to 10^{-5} torr in 10 sec. Assume that there is no gas source to the chamber during shutdown and use the properties of H_2.

11.3. What total area of cryopanels is required to provide the pumping speed in Problem 11.2?

11.4. Estimate the tritium permeation rate through a first wall of area 500 m² and thickness 0.5 cm with a plasma side tritium pressure of 10^{-2} torr and a coolant side tritium pressure of 10^{-6} torr. Consider a stainless-steel wall operating at 450 °C and a vanadium wall operating at 600 °C.

11.5. Calculate the doubling time for a 500 MW$_{th}$ D-T fusion reactor with a tritium breeding ratio of $TBR = 1.1$ and an initial external tritium inventory of $I_0 = 10$ kg. The mean residence time for tritium in the blanket is $\tau_B = 10$ hr, the mean times against unrecoverable tritium loss are $\tau_{LB} = 100$ hr and $\tau_{L\ ex} = 10$ yr, and the permeation loss through the first wall is 1% of the fusion rate ($\varepsilon = 0.01$).

11.6. Calculate the buildup of the tritium inventory in the blanket of Problem 11.5 from startup until it reaches its asymptotic value.

12
Fusion Reactors

The physical characteristics of future fusion reactors will be constrained, and therefore determined, by limits on the underlying plasma physics and fusion technology. In this chapter we first discuss these plasma physics and fusion technology limits and then review the projected physical characteristics of future fusion reactors and neutron sources that are determined by these limits. Since the plasma physics understanding is sufficiently well developed to make such an exercise meaningful only for the tokamak confinement concept, we will limit detailed discussion of these considerations to the tokamak.

12.1
Plasma Physics and Engineering Constraints

We have discussed a number of limits – on confinement, on pressure, on density, and so on – that are imposed by various plasma phenomena. Some of these limits (e.g., MHD instabilities) are reasonably well understood at a fundamental level, while others (e.g., energy confinement) exist more as empirical limits of broad generality. We will use the nomenclature of Figure 12.1 to discuss the effect of the various limits on the dimensions of a tokamak.

12.1.1
Confinement

The requirement to achieve a given level of energy confinement time, as needed to satisfy the plasma power balance, imposes a constraint on the set of physics and engineering parameters that determine energy confinement. At present, this set of parameters and the form of the constraint is best represented by an empirical confinement scaling law. Using the ITER IPB98(y,2) scaling

$$\tau_E = H_H \tau_E^{IPB98(y,2)}$$

Fusion: Second, Completely Revised and Enlarged edition. Weston M. Stacey
Copyright © 2010 Wiley-VCH Verlag GmbH & Co. KGaA, Weinheim
ISBN: 978-3-527-40967-9

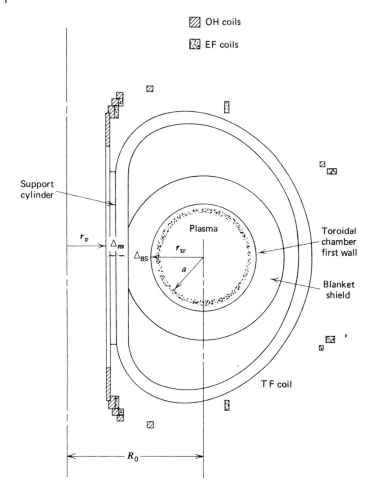

Figure 12.1 Nomenclature of tokamak configuration.

where

$$\tau_E^{IPB98(y.2)} = 0.0562 I^{0.93} B^{0.15} P^{-0.69} \bar{n}_{e20}^{0.41}$$
$$\times M^{0.19} R^{1.97} A^{-0.58} \kappa^{0.78},$$

imposes a constraint among plasma current I_p(MA), magnetic field B(T), heating power P(MW), line-average electron density $\bar{n}_e(10^{20}/m^3)$, plasma ion mass M(amu), major radius R(m), minor radius a(m) via $A = R/a$, and plasma elongation κ. H_H is the confinement enhancement factor relative to the above correlation of H-mode data (for which $H_H = 1$). The objective of current confinement research, in addition to understanding the phenomena determining confinement, is to increase H_H above unity. It is encouraging that $H_H > 1$ has been obtained under various conditions on several tokamaks. This equation indicates that the major factors determining confinement are the plasma current and the size of the plasma.

There is a threshold power crossing the last closed flux surface (LCFS) that is required for access to H-mode confinement. While there are some theoretical models for this power, this threshold power must at present be taken from the empirical correlation

$$P_{LH}(MW) = (2.84/M) B^{0.82} \bar{n}_{e20}^{0.58} Ra^{0.81}$$

which defines a further constraint on magnetic field and plasma size. Providing a theoretical understanding of this power threshold is an active area of plasma physics research.

12.1.2
Density Limit

The fusion power in a 50–50 D-T plasma is given by

$$P_{DT} = \frac{1}{4} n^2 \langle \sigma v \rangle U (2\pi Ra^2 \kappa)$$

where $U = 3.5$ MeV for the alpha heating of the plasma and $U = 17.6$ MeV for the total fusion power. Clearly, achieving high power density requires achieving high particle density.

The empirical Greenwald density limit

$$\bar{n}_{e20} \leq \frac{I_p(MA)}{\pi a^2}$$

provides a reasonable bound on a large amount of experimental data, although there are many discharges that exceed this limit by up to a factor of 2, and there are many discharges that are limited to densities that are only a fraction of the Greenwald limit. This simple expression, which is good to within roughly a factor of 2, indicates that high current density is important for achieving high plasma density, thus defining a further constraint among plasma current, plasma size and fusion power.

12.1.3
Beta Limit

The MHD stability limit on the plasma pressure

$$\beta_c \equiv \frac{\langle n_e T_e + n_i T_i + p_\alpha \rangle}{\frac{B^2}{2\mu_0}} \leq \beta_N \frac{I(MA)}{aB} = const \times l_i \frac{I}{aB}$$

where the internal inductance is given by

$$l_i = \ln(1.65 + 0.89(q_{95} - 1))$$

bounds a wide range of discharges from several tokamaks. Here, as elsewhere in the text, q_{95} refers to the safety factor evaluated on the flux surface that encloses 95% of the plasma volume.

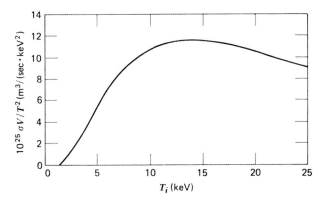

Figure 12.2 Fusion reaction parameter $\langle \sigma v/T^2 \rangle$ for D-T.

In a beta-limited tokamak, this equation can be solved for $nT \sim \beta_t B^2$ and this result can be used to write the above expression for the fusion power as

$$P_{DT} \sim \beta_t^2 B^4 \frac{\langle \sigma v \rangle}{T^2}(2\pi R\pi a^2 \kappa)$$

$$\sim \beta_N^2 I^2 B^2 \frac{\langle \sigma v \rangle}{T^2}(2\pi R\pi \kappa)$$

The quantity $\langle \sigma v \rangle \sim T^2$ at the lower temperatures of the thermonuclear regime that are of the greatest practical interest, so that $\langle \sigma v \rangle/T^2$ has a broad peak in the temperature range $10 < T < 20\,keV$, as shown in Figure 12.2.

The above relation for the fusion power highlights the importance of achieving high values of β_N, I and B in order to achieve high plasma power density and also defines a constraint among these variables and plasma size for a given fusion power level. The achievement of high values of β_N is a major focus of plasma physics research.

If the alpha particle "ash" is not exhausted from the plasma but allowed to build up, the alpha pressure, p_α, buildup will reduce the fraction of β_t that can be used to confine the D-T ions, and P_{DT} will be correspondingly reduced. Thus, alpha exhaust is an important research issue.

12.1.4
Kink Stability Limit

The kink stability limit can be roughly characterized by the requirement

$$q_{95} = \frac{5a^2 B}{RI}\left(\frac{1 + \kappa^2(1 + 2\delta^2 - 1.2\delta^3)}{2}\right)\frac{\left(1.17 - \dfrac{0.65}{A}\right)}{\left(1 - \dfrac{1}{A^2}\right)^2} \geq 3.$$

where δ is the plasma triangularity. This limit imposes a constraint on the allowable combination of B, I, R and a.

12.1.5
Startup Inductive Volt-seconds

The usual practice is to provide enough inductive volt-seconds through transformer action of the poloidal coil system to start up and at least partially maintain the plasma current for the entire discharge. While maintenance of the discharge largely by non-inductive current drive is envisioned for future tokamaks, it will always be prudent to have a capability at least for inductive startup. The startup volt-second requirement consists of a component to induce the plasma current in the absence of resistive losses

$$(\Delta\Phi)_{ind} = IL_p$$

plus a component to overcome resistive losses during startup

$$(\Delta\Phi)_{res} = C_{Ejima}\mu_0 RI$$

where the Ejima coefficient is about 0.4 and the plasma inductance is given by

$$L_p = \mu_0 R\left[\ln\left(\frac{8R}{a\sqrt{\kappa}}\right) + \frac{l_i}{2} - 2\right]$$

With reference to Figure 12.1, all of the poloidal coils – the central solenoid (CS) and other "ohmic heating" coils and the equilibrium field (EF) coils – can, depending on the design, provide a flux linkage with the plasma and thus contribute to providing volt-seconds. Normally, the majority of the volt-seconds are provided by the CS.

The total required inductive volt-seconds to maintain the plasma discharge after startup will be determined by a trade-off involving the required length of the burn pulse (availability requirement and fatigue limits), the practically achievable boot-strap current, the amount of power required for non-inductive current drive, and the overall design of the coil systems. These volt-seconds plus the startup volt-seconds discussed above must be supplied by the poloidal coil system, and some part of this, $\Delta\Phi_{CS}$, must be provided by the central solenoid.

Providing $\Delta\Phi_{CS}$ volt-seconds requires

$$(\Delta\Phi)_{CS} = \pi B_{OH} r_v^2 \left[1 + \frac{\Delta_{OH}}{r_v} + \frac{1}{3}\left(\frac{\Delta_{OH}}{r_v}\right)^2\right]$$

where B_{OH} is the maximum field in the CS, r_v is the flux core radius shown in Figure 12.1, and Δ_{OH} is the thickness of CS (which contributes to the "magnet" thickness Δ_m indicated in Figure 12.1). The CS thickness Δ_{OH} must be sufficient so that the tensile stress in the CS

$$\sigma_{CS} = \frac{B_{OH}^2}{2\mu_0}\left(\frac{r_v}{\Delta_{OH}} + \frac{1}{3}\right) \le S_m,$$

satisfies the American Society of Mechanical Engineers (ASME) Code; that is, $S_m \leq \min$ [1/3 ultimate stress, 2/3 yield stress].

12.1.6
Non-Inductive Current Drive

The plasma current during the discharge in present tokamaks is maintained at least in part inductively against resistive losses. However, because of implications for fatigue limits and availability, this is not an attractive option for future tokamak reactors. So, current maintenance during the discharge by a combination of non-inductive current drive and bootstrap current is envisioned for future tokamak reactors.

The current drive efficiency, $\eta_{CD} \equiv I_{CD}(MA)/P_{CD}(MW)$, achieved to date on a number of tokamaks is given in Figure 5.8.

Neutral beam and electromagnetic wave current drive favor high electron temperature and low density to achieve high efficiencies. For fast wave current drive, a useful formula is

$$\gamma_{FW} \equiv R_0 n_{e20} \eta_{CD} = 0.062 T_e (keV)^{0.56}$$

The present current drive efficiencies are modest, implying the need for both improvement in the efficiency of non-inductive current drive and the achievement of a high bootstrap current fraction – both items are being actively investigated.

12.1.7
Bootstrap Current

The bootstrap current fraction of the total current is given by

$$f_{bs} = C_{BS} \left(\sqrt{\varepsilon} \beta_p \right)^{1.3},$$

where

$$C_{BS} = 1.32 - 0.235 q_{95} + 0.0185 q_{95}^2$$

and

$$\beta_p = \beta_t (B/B_p)^2, \quad B_p = \frac{I_p(MA)}{5a\sqrt{\dfrac{1+\kappa^2}{2}}}$$

Achievement of a high bootstrap current fraction clearly depends on achieving a high plasma pressure.

12.1.8
Toroidal Field Magnets

Use of a large magnetic field, *B*, in the plasma was shown above to be of value in achieving favorable values of several plasma parameters. However, electromagnetic forces in the toroidal field (TF) coils scale like $F \sim I_{TF} \times B_{TF} \sim B_{TF}^2$, so high magnetic fields require a lot of structure in the coils in order to remain within stress limits, which increases the magnet thickness Δ_m in Figure 12.1.

The "centering" force (directed radially inward towards the center of the tokamak) of the TF coil system is

$$ F_R = \frac{\mu_0 N I_{TF}^2}{2} \left[1 - \frac{1}{\sqrt{(1-\varepsilon_p^2)}} \right], $$

where N is the number of TF coils, I_{TF} is the current flowing in each TF coil, and $\varepsilon = R_{bore}/R$ (where R_{bore} is the bore radius of the TF coil and R is the major radius from the central axis of the tokamak to the center of the TF coil). This centering force must be reacted either by a support cylinder (see Figure 12.1) or by wedging the TF coils (like a stone arch).

The tensile force in the TF coil is

$$ F_T = \frac{1}{2} \frac{\mu_0 N I_{TF}^2}{4\pi} \ln\left(\frac{1+\varepsilon_p}{1-\varepsilon_p}\right), $$

and the corresponding tensile (hoop) stress is

$$ \sigma_t = F_T / A_{TF} $$

where A_{TF} is the cross-sectional area of the TF coil (excluding coolant and insulator). There is a bending stress due to the interaction among adjacent TF coils (for off-normal conditions) that is also $\sim B^2{}_{TF}$. The ASME code requires

$$ \sigma_t + \sigma_{bend} \leq 1.5 \, S_m $$

where S_m is defined in Section 12.1.5. Since

$$ B_{TF} = \frac{\mu_0 N I_f}{2\pi R} = \frac{\mu_0 N I_f}{2\pi (r_v + \Delta_m)} $$

the centering, tensile and bending forces cause the contribution of the TF coils to the magnet thickness Δ_m to scale $\sim B^2{}_{TF}$.

The toroidal field in the plasma, B, is related to the peak toroidal field in the TF coil, B_{TF}, which is located at $(r_v + \Delta_m)$ by

$$ B = B_{TF}\left(1 - \frac{r_w + \Delta_{BS}}{R}\right) = B_{TF}\frac{(r_v + \Delta_m)}{R} $$

which follows from the $B \sim 1/R$ scaling dictated by Ampere's law taken around a toroidal loop at major radius R.

12.1.9
Blanket and Shield

The thickness of the blanket plus shield, Δ_{BS}, must be sufficient to: (i) include a lithium-containing region for producing tritium by neutron capture in order to replace the tritium consumed in the fusion reactions; (ii) transform >95% of the fusion energy which is in the form of fast neutrons into heat energy by neutron scattering reactions and remove it for conversion to electricity; and (iii) shield the sensitive superconducting magnets from neutron irradiation damage and heating by neutrons and capture gammas. Studies indicate that $1.0 < \Delta_{BS} < 1.5$ m is required to accomplish these objectives.

12.1.10
Plasma Facing Component Heat Fluxes

There are limits on the allowable peak heat flux to the plasma facing components surface arising from several different phenomena. The most limiting of these constraints places a lower limit on the required surface area needed to handle a given amount of plasma exhaust power, hence to handle a given fusion alpha power plus auxiliary heating power level.

12.1.10.1 Heat Fluxes
The peak heat flux to the "first wall" of the surrounding plasma chamber is

$$q''_{FW} = \frac{(0.2P_{fus} + P_{aux})(1 - f_{div})\hat{f}_{fw}}{(2\pi R)\left[2\pi a \sqrt{\frac{1}{2}(1 + \kappa^2)}\right](1 - \varepsilon_{div})}$$

where $f_{div} = f_{\pm}$ is the fraction of the plasma power that goes to the divertor, as opposed to the first wall ($f_{div} \geq 0.5$ is a representative range), and $\hat{f}_{fw}$ is the first wall heat flux peaking factor ($\hat{f}_{fw} \leq 2$ is representative), and ε_{div} is the fraction of the area around the plasma that is occupied by the divertor.

The divertor is represented here as a fraction ε_{div} of the first wall surface area. The divertor target plate is represented as a toroidal strip of thickness Δ_{div} circling the torus at the major radius, R. The peak heat flux on the divertor target plate can be written

$$q''_{DP} = \frac{(0.2P_{fus} + P_{aux})f_{div}f_{dp}\hat{f}_{dp}}{\Delta_{div}2\pi R}$$

where f_{dp} is the fraction of the exhaust power going to the divertor that ends up on the divertor plate, as opposed to being radiated to the divertor channel walls ($f_{dp} \approx 0.3$–0.9 is representative), and $\hat{f}_{dp}$ is the divertor target plate heat flux peaking factor ($\hat{f}_{dp} \approx 5$–10 is representative).

12.1.10.2 Stress Limits
Using a tube bank model of the plasma facing components, as indicated in Figure 12.3, the primary stress in the coolant tubes is given by

$$\sigma_p = \frac{(p_c + p_{dis})r_c}{t_c} \leq S_m$$

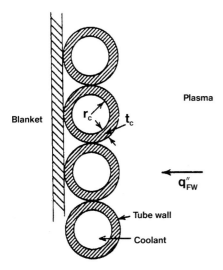

Figure 12.3 Tube bank model of plasma facing components.

where p_c is the coolant pressure, r_c is the coolant tube radius, t_c is the coolant tube thickness, and $p_{dis} \simeq B_\theta^2/2\mu_0$ is the disruption pressure.

The thermal bending stress due to a temperature gradient across the coolant tube wall is

$$\sigma_{th} = \frac{\alpha E\left(f_{pfc}q''t_c + \frac{1}{2}q'''t_c^2\right)}{2\kappa(1-\nu)}$$

where f_{pfc} is an adjustable parameter normalized to more detailed modeling ($f_{pfc} \approx 4/3$ is a representative value), q'' is the surface heat flux from the plasma, q''' is the volumetric heat source due to neutrons, α is the coefficient of expansion, E is Young's modulus, κ is the thermal conductivity, and ν is Poisson's ratio.

The ASME Code specifies a maximum combination of primary plus bending stress

$$\sigma_{th} + \sigma_p \leq 3S_m$$

Since both σ_{th} and σ_p depend on the q'', this equation specifies a stress limit on the maximum allowable heat flux to the plasma facing component, q''_{stress}.

12.1.10.3 Temperature Limit

The maximum surface temperature of the plasma facing component is usually constrained to remain below some maximum value, T_s^{max}, usually the melting temperature. The surface temperature constraint can be written

$$T_s = T_c + f_{pfc}q''_T\left(\frac{t_c}{\kappa} + \frac{1}{h}\right) + \frac{q'''t_c^2}{2\kappa} \leq T_s^{max}$$

where T_c is the coolant temperature and h is the surface heat transfer "film" coefficient. This relation defines a maximum allowable heat flux, q''_T.

12.1.10.4 Fatigue Limit

When the plasma facing components are subjected to a cyclic loading (e.g., variation in thermal stress due to a repetitively pulsed mode of tokamak operation) they will fail after a mean number of cycles, $N_D(\varepsilon)$, which depends on the material and the cyclic strain, ε, produced by the cyclic loading (e.g., thermal stress). Using the above expression for the thermal stress, which is related to the thermal strain, the maximum allowable heat flux can be written

$$q''_{fat} = \frac{2\kappa(1-\nu)\,\varepsilon_{max}(N_D)}{f_{pfc}\,\alpha} \frac{1}{t_c} - \frac{1}{2}q'''\,t_c$$

where ε_{max} is the maximum allowable strain for N_D cycles (from fatigue curves).

We only discuss fatigue here, but the phenomenon of growth of microscopic defects under cyclic loading also takes place and constitutes a similar limit on the surface heat flux.

12.1.10.5 Heat Flux Limit

The heat flux limit on the plasma facing components is just the most limiting of the above three maximum allowable heat fluxes $q''_{max} = \max\left\{q''_{th}, q''_T, q''_{fat}\right\}$.

Although present tokamaks are not limited by the surface heat flux to plasma facing components, this will be an issue for future tokamaks with more auxiliary heating power and fusion alpha power to exhaust. The ability to take advantage of anticipated advances in plasma confinement and β-limits to achieve compact, high power density tokamaks may be thwarted by heat flux limits unless advanced materials that can handle higher surface heat loads are also developed.

12.1.11
Radiation Damage to Plasma Facing Components

The first wall average 14 MeV neutron power flux is

$$\Gamma_n = 0.8 P_{fus} \left/ 2\pi R \left[2\pi a \sqrt{\frac{1}{2}(1+\kappa^2)} \right] \right.$$

and the first wall peak 14 MeV neutron power flux is

$$\hat{\Gamma}_n = \Gamma_n \hat{f}_n \sqrt{\frac{1}{2}(1+\kappa^2)}$$

where $\hat{f}_n$ is the neutron poloidal peaking factor (typically 1.1–1.3).

The 14 MeV fusion neutrons incident on the plasma facing components displace atoms from the lattice and transmute atoms of the wall material. These microscopic changes manifest themselves as swelling, loss of ductility, reduction of the number of cycles to fatigue failure for a given strain or in the allowable strain for given number of cycles, and other macroscopic changes in materials properties that lead to failure of the plasma facing components to perform their intended function. This radiation damage does not directly limit the size or other parameters of a tokamak, but it does determine the radiation damage lifetime of the plasma facing components. If this lifetime is too short and it is necessary to replace the component too frequently, then the design is impractical. One direct effect on the design of a tokamak is the reduction in the allowable strain to achieve a given number of cycles before fatigue failure, which requires an increase of the coolant tube thickness that in turn causes a reduction in the stress limited allowable surface heat flux.

Radiation damage limits are conveniently expressed in terms of the energy fluence of 14 MeV neutrons, in units of MW-yr/m^2. The average neutron flux is Γ_n (MW/m^2) $= 0.8\, P_{fus}$(MW)$/A_{wall}$(m^2), and the fluence is Φ(MW-yr/m^2) $= \Gamma_n$ (MW/m^2) $\times \Delta t$ (yr). A neutron flux of 1 MW/m^2 corresponds to 4.48×10^{17} 14 Mev neutrons per m^2 per second. The only structural material that is presently qualified for building a tokamak reactor is austenitic stainless steel, which has a radiation damage lifetime of about 4 MW-yr/m^2, which would be less than one effective full power year for the type of tokamak reactors presently envisioned (see Sections 12.4 and 12.5). Ferritic steels and vanadium alloys are under development which promise to have much greater radiation damage lifetimes. Unless the development of an advanced material with a radiation damage lifetime greater than $\approx$10–20 MW-yr/m^2 is developed, it may prove impossible to take advantage of the anticipated advances in plasma confinement and beta limits to achieve a practical compact, high power density fusion reactor.

12.2
International Tokamak Program

The international tokamak program began with T-1 at the Kurchatov Institute in Moscow in 1957. When the then surprisingly high temperatures found in T-3 were reported at an international meeting and subsequently confirmed by a British team, the world fusion community put other confinement concepts aside and turned its attention to the tokamak. In the intervening 50 years about 60 experiments (see Table 12.1) in the USSR, the USA, Europe, Japan and other countries have developed our present understanding of tokamak physics and advanced the fusion triple product parameter $n T \tau$ by about 5 orders of magnitude. While numerous other magnetic confinement concepts have also been investigated over this period, many of them fell by the wayside relative to the tokamak because of inability to overcome scientific or engineering problems, and of those remaining none are nearly as developed as the tokamak.

12.3
The International Thermonuclear Experimental Reactor (ITER)

In 1979, the USA, USSR, Europe (EC) and Japan came together under the auspices of the International Atomic Energy Agency in a series of International Tokamak Reactor (INTOR) Workshops that spanned the following decade to assess the readiness of the tokamak: (i) to move forward to a "burning plasma", or experimental reactor, phase; (ii) to identify the physical characteristics and required R&D for such a device; and (iii) to perform supporting feasibility and conceptual design analyses. Based on the

Table 12.1 Tokamaks.

First Year	Device	R(m)	a(m)	B(T)	I (MA)	Divertor	NBI P(MW)	ICRH P(MW)	LHR P(MW)	ECR P(MW)
1957	T-1[1]	0.63	0.13	1.0	0.04					
1962	T-3[1]	1.0	0.12	2.5	0.06					
1963	TM-3[1]	0.4	0.08	4.0	0.11					
1968	LT-1[11]	0.4	0.10	1.0	0.04					
1970	ST[2]	1.09	0.14	4.4	0.13					
	T-4[1]	1.0	0.17	5.0	0.24					
1971	Ormak[2]	0.8	0.23	1.8	0.20		0.34			
	T6[1]	0.7	.25	1.5	0.22					
	TumanII[1]	0.4	0.08	2.0	0.05					
1972	ATC[2]	0.88–0.35	0.17–0.11	2.0–5.0	0.11–0.28		0.1	0.16		0.2
	Cleo[3]	0.9	0.18	2.0	0.12		0.4			0.4
	DII[2]	0.60	0.10	0.95	0.21					
	JFT-2[8]	0.9	0.25	1.8	0.23		1.5	1.0	0.3	0.2
	T-12[1]	0.36	0.08	1.0	0.03					
	TO-1[1]	0.6	0.13	1.5	0.07					
1973	AlcatorA[2]	0.54	0.1	10.0	0.31			0.1	0.1	
	Pulsator[2]	0.7	0.12	2.7	0.09					
	TFR400[4]	0.98	0.20	6.0	0.41		0.7			
1974	DIVA[8]	0.6	0.10	2.0	0.06	x				
	Petula[4]	0.72	0.16	2.7	0.16				0.5	
	Tosca[3]	0.3	0.09	0.5	0.02					0.2
1975	DITE[3]	1.17	0.26	2.7	0.26	x	2.4			
	FT[5]	0.83	0.20	10.0	0.80				1.0	
	PLT[2]	1.3	0.40	3.5	0.72		3.0	5.0	1.0	

Year	Device									
	T-10[1]	1.5	0.37	4.5	0.68					1.0
	T-11[1]	0.7	0.22	1.5	0.17					
1976	JIPPT2[8]	0.91	0.17	3.0	0.16		0.7		0.2	
	Microtor[2]	0.3	0.10	2.5	0.14		0.1			
	TNT-A[8]	0.4	0.10	0.44	0.02		0.5			
1977	ISX-A[2]	0.92	0.26	1.8	0.22					
	Macrator[2]	0.9	0.40	0.4	0.12			0.5		
1978	ISX-B[2]	0.93	0.27	1.8	0.24		2.5	1.5		0.2
	TFR600[4]	0.98	0.20	6.0	0.41					0.6
	TUMANIII[1]	0.55	0.15	3.0	0.20					
	Versator[2]	0.4	0.13	1.5	0.11					0.1
1979	AlcatorC[2]	0.64	0.17	12.0	0.90	×			0.1	
	PDX[2]	1.4	0.45	2.5	0.60	×	7.0		4.0	
1980	ASDEX[6]	1.54	0.40	3.0	0.52	×	4.5	3.0	2.0	
	DIII[2]	1.45	0.45	2.6	0.61	×	7.0			2.0
	TCA[13]	0.61	0.18	1.5	0.17					0.4
	LT-4[11]	0.5	0.10	3.0	0.10					
1981	TEXT[2]	1.0	0.27	2.8	0.34	×				0.6
	T-7[1]	1.22	0.31	3.0	0.39				0.25	
1982	TFTR[2]	2.4	0.80	5.0	2.2		40.0	11.0		
1983	HT-6B[9]	0.45	0.12	0.75	0.04				0.1	0.1
	JET[7]	3.0	1.25	3.5	7.0	×	20.0	20.0	7.0	
	TEXTOR[6]	1.75	0.46	3.0	0.6		4.0	4.0		
1985	HT-6M[9]	0.65	0.2	1.5	0.15			1.0	0.15	0.2
	JT60[8]	3.0	0.95	4.5	2.3	×	20.0	2.5	7.5	
1986	DIII-D[2]	1.67	0.67	2.1		×	20.0	4.4		2.0
1988	Tore Supra[4]	2.37	0.80	4.5	2.0	ergodic	1.7	12.0	8.0	2.0
	T-15[1]	2.40	0.70	3.6	1.0		1.0			2.0

(Continued)

Table 12.1 (Continued)

First Year	Device	R(m)	a(m)	B(T)	I (MA)	Divertor	NBI P(MW)	ICRH P(MW)	LHR P(MW)	ECR P(MW)
1989	COMPASS[3]	0.56	0.21	2.1	0.28	x			0.6	2.0
	RTP[12]	0.72	0.16	2.5	0.16					0.9
	ADITYA[14]	0.75	0.25	1.5	0.25			0.2		0.2
1990	FT-U[5]	0.93	0.30	8.0	1.3			2.0	4.0	1.0
1991	ASDEX-U[6]	1.65	0.50	3.9	1.4	x	10.0	6.0		0.5
	JT60-U[8]	3.4	1.1	4.2	5.0	x	40.0	7.0	8.0	
1992	START[3]	0.2–0.3	0.15–.24	0.6	0.12	x				0.2
	TCV[13]	0.88	0.25–0.7	1.4	1.2					4.5
	KAIST[10]	0.53	0.14	0.5	0.12					
1993	AlcatorCM[2]	0.67	0.22	9.0	1.1	x		4.0		
	ET[2]	5.0	1.0	.25–1.	0.045		1.0	2.0		
2006	EAST[9]	1.7	0.4	3.5	1.0	x		3.0	3.5	0.5
2009	KSTAR[10]	1.8	0.5	3.5	2.0	x	14.0	6.0	3.0	4.0
	SST-1[14]	1.10	0.2	3.0	0.22	x	0.8	1.5	1.0	0.2
	T-15[1]	2.43	0.42	3.5	1.0	x	9.0		4.0	7.0
2017	ITER	6.2	2.0	5.3	15.0	x	50	40	40	20

Country: [1]USSR, [2]US, [3]UK, [4]France, [5]Italy, [6]Germany, [7]European, [8]Japan, [9]China, [10]Korea, [11]Australia, [12]Netherlands, [13]Switzerland, [14]India.

(Notation: R = major radius, a = minor radius, B = toroidal magnetic field. I = plasma current, NBI = neutral beam injection, ICRH = ion cyclotron resonance heating, LHR = lower hybrid resonance heating, ECR = electron cyclotron resonance heating. Units: m = meters, T = Tesla, MA = million amperes, MW = million watts).

positive results of this work, Secretary Gorbachev proposed to President Reagan at the Geneva Summit in 1985 the collaborative design, construction and operation of such an INTOR device. This proposal subsequently led to the formation of the International Thermonuclear Experimental Reactor (ITER) project involving the same Parties (and since expanding to include also China, India and South Korea).

As of 2009, the design of ITER had been developed, reviewed many times by numerous international bodies, and revised accordingly; the supporting R&D had been completed; the governments of the Parties to ITER had agreed to precede with a joint international project; a site had been selected in Cadarache France; a project team had begun to assemble; and the construction of the ITER device had begun. Initial operation is scheduled within a decade.

The ITER device is illustrated in Figure 12.4, and the major parameters are given in Table 12.2. The purpose of ITER is to establish the scientific and technological feasibility of fusion. To this end, ITER is intended to: (i) investigate moderate-Q_p burning plasma conditions, with an ultimate goal of achieving steady-state plasma operation; (ii) to integrate into a single fusion system reactor-relevant nuclear and plasma support technologies and a reactor-relevant fusion plasma; (iii) to test high heat flux and fusion nuclear components; and (iv) to demonstrate the safety and environmental acceptability of fusion. In order to achieve these objectives, ITER is designed to achieve $Q_p \geq 10$, an inductive burn pulse of $\geq 400\,\mathrm{s}$, a fusion power of $500\,\mathrm{MW}$, an average first wall neutron flux $\geq 0.5\,\mathrm{MW/m^2}$ and a lifetime fluence $\geq 0.3\,\mathrm{MW\text{-}yr/m^2}$, with the ultimate objective of steady-state operation at $Q_p \geq 5$.

The ITER magnets are superconducting, with Nb_3Sn conductor for the toroidal fields coils and central solenoid, and with NbTi conductor for the poloidal ring coils. The vacuum vessel/first-wall is water-cooled copper, coated on the plasma side with beryllium, bonded onto an austenitic stainless steel structure. The divertor target plate has a similar configuration but with tungsten coating, except near the peak heating locations where a carbon fibre composite is used.

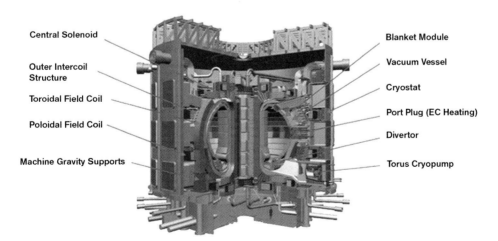

Figure 12.4   ITER Design-Main Features (Reproduced with permission from the ITER website.)

Table 12.2 ITER parameters.

Parameter	
P_{fus} (MW$_{th}$)	410
Paux (MW)	40
$Q_p = P_{fus}/P_{aux}$	10
P_{lcfs}/P_{L-H}	1.6
R(m)	6.2
a(m)	2.0
B(T)	5.3
I (MA)	15.0
n(10^{20}/m^3)	1.0
T_e/T_i (keV)	8.9/8.1
H_H(enhancement of H-mode)	1.0
τ_E(s)	3.7
β_n	1.8
q_{95}	3.0
κ	1.8
δ	0.4
n/n$_{greenwald}$	0.85
q_{rw} (MW/m^2)	0.15
Γ_n (MW/m^2)	0.5

As an example of but one of the thousands of extensive analyses that have been carried out in support and evaluation of the ITER design. Figure 12.5 depicts the range of operating conditions for which ITER can achieve the $Q_p \geq 10$ design objective. Perhaps the most important uncertainties that could affect ITER performance are the achievable energy confinement and plasma density. The solid lines in Figure 12.5 indicate the value of the energy confinement enhancement factor, relative to the empirical fit of H-mode tokamak experimental confinement times, that would be required to achieve $Q_p = 10$, for various values of the plasma density as a fraction of the Greenwald density limit. The upper dashed lines indicate the contours of the normalized β_N, which are generally below the limiting values for currently operating tokamaks. The lower dotted line indicates the contour for which the non-radiative power crossing the separatrix is 30% greater than the threshold value for access to H-mode. It is clear that ITER is conservatively designed to achieve the $Q_p \geq 10$ design objective.

12.4
Future Tokamak Reactors

The H-mode confinement scaling, the beta limit, the current drive efficiencies, and so on which can be routinely achieved today are known as the "conventional" tokamak database, which forms the basis for the ITER design. A future tokamak reactor designed on relatively modest extensions of this "ITER database" is

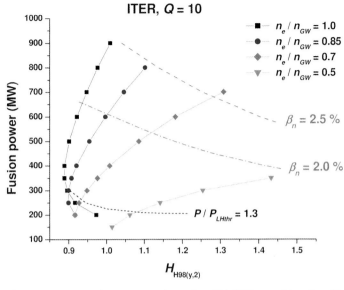

Figure 12.5 Range of operating conditions for which ITER can achieve $Q_p = 10$ (Reproduced with permission from the Georgia Tech FRC website.)

described by the parameters in the first column (labeled "ARIES-I") of Table 12.3. The parameters shown in the second column (labeled "High Field ARIES-I" indicate the improvements that could be obtained if it was possible to extend the toroidal field strength at the coil from 16 to 19 T (with a proportional increase of the

Table 12.3 Major parameters of 1000-MWe future tokamak power plants.

	First stability		Reverse-shear	
	ARIES-I	High-field: ARIES-I	ARIES-RS	ARIES-AT
Major radius (m)	8.0	6.75	5.5	5.2
Plasma aspect ratio	4.5	4.5	4.0	4.0
Plasma elongation, κ	1.8	1.8	1.9	2.1
β (%) (β_n)	2 (2.9)	2 (3.0)	5 (4.8)	9.2 (5.4)
Peak field at the coil (T)	16	19	16	11.5
On-axis field (T)	9.0	11	8	5.8
Neutron wail load (MW/m^2)	1.5	2.5	4	3.3
ITER89-P multiplier	1.7	1.9	2.3	2.0
Plasma current (MA)	12.6	10	11.3	13
Bootstrap current fraction	0.57	0.57	0.88	0.91
Current-driver power (MW)	237	202	80	35
Recirculating power fraction	0.29	0.28	0.17	0.14
Thermal efficiency	0.46	0.49	0.46	0.59
Cost of electricity (¢/kWh)	10	8.2	7.5	5

field in the plasma), to extend the plasma confinement by 11% and to increase the limiting β_N from 2.9 to 3.0.

Much of the emphasis in the tokamak R&D programs is on extending this "conventional" performance capability to create an advanced tokamak physics database for the design of future tokamak reactors. In order to take advantage of such an improved physics database to improve the design of tokamak reactors, it will also be necessary to develop an advanced engineering technology database, and some thought has been given to this. The last two columns in Table 12.3 (labeled "Reverse Shear") take advantage of anticipated physics advances that would allow operation at higher β_N with improved plasma confinement to achieve a more compact design. Note that a higher neutron wall load is associated with these more compact designs, which would necessitate the development of more radiation resistant first wall structural materials in order to avoid the more frequent first-wall replacements that would otherwise be associated with the larger neutron wall load.

The ARIES-AT design for an advanced tokamak power reactor is depicted in Figure 12.6.

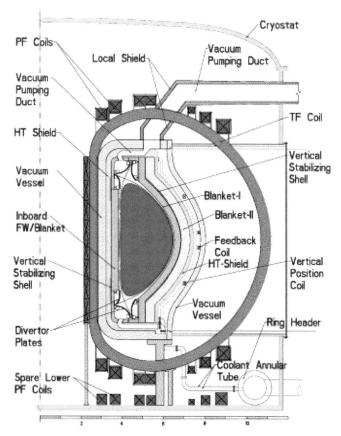

Figure 12.6 The ARIES-AT Advanced Tokamak Reactor Design (Reproduced with permission from the ARIES website.)

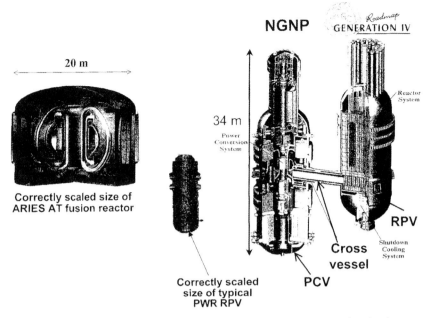

Figure 12.7 Comparison of scale of fusion and fission power reactors (Reproduced with permission from the ARIES website.)

It provides a useful perspective to compare the overall size of the ARIES-AT fusion reactor with a fission reactor that produces a comparable power, which is done in Figure 12.7. The volume inside of the fusion reactor pressure vessel is considerably larger than the volume within the pressure vessel of a conventional Presssurized Water Reactor (PWR), but is comparable to the volume within the pressure vessel enclosing the reactor system for the gas-cooled "next geneneration nuclear plant" (NGNP) envisioned in the Generation IV development plan.

12.5
Tokamak Demo Reactor

Following the operation of ITER, and before the construction of an advanced "commercial" tokamak reactor like ARIES-AT, it is anticipated that a generation of Demonstration Reactors would be built for the purpose of demonstrating reactor-level performance (e.g., availability >50%, tritium self-sufficiency) and the potential for economic competitiveness with other energy sources. Assuming that advanced tokamak physics is developed in parallel with ITER and tested to a limited extent on ITER, and further assuming that an advanced structural material capable of handling high surface heat loads and high neutron fluxes is developed in parallel with ITER, one possible set of parameters for an AT-DEMO is compared with the ARIES-AT parameters in Table 12.4.

Table 12.4 Possible parameters of future advanced tokamak reactors.

Parameter	AT-DEMO	ARIES-AT
P (MW$_e$)	400	1000
R(m)	5.4	5.2
B(T)	6.0	5.8
I (MA)	9.4	13
H$_{89}$-P	2.0	2.0
β$_n$	4.2	5.4
q$_{95}$	4.0	3.7
f$_{bs}$[a)	0.7	0.9
κ	2.0	2.1
Γ$_n$ (MW/m^2)	2.0 (average)	3.3 (max.)

a) bootstrap current fraction.

12.6
Tokamak Neutron Source

While the development of magnetic fusion has been primarily motivated by the promise of a limitless and environmentally benign source of electrical power, there are other potential "non-electrical" applications of fusion as a source of neutrons, some of which may be possible to undertake at an earlier stage in the development of fusion science and technology. The use of fusion neutron sources to produce hydrogen for a hydrogen economy, or tritium for national security purposes, or radioisotopes for medical and industrial applications, or neutrons for materials and fundamental research, have all been suggested.

It has also been suggested that a fusion neutron source could be used with a sub-critical nuclear reactor to "breed" fissionable nuclei (^{239}Pu or ^{233}U) from non-fissionable (with "thermal" neutrons) nuclei (^{238}U or ^{232}Th), to "dispose" of surplus weapons-grade plutonium by transmuting enough of the ^{239}Pu into ^{240}Pu to make the mixture unsuitable for weapons, or to transmute (fission) the plutonium and higher actinides remaining in spent nuclear fuel instead of storing that material for millions of years in high level waste repositories. All of these "nuclear" missions have in common that most of the neutrons are provided by the fission reactor and that the purpose of the fusion neutrons is to help maintain the neutron chain reaction in the subcritical reactors, thus the performance requirements (for high beta, high density, high confinement, etc.) for the fusion device are much less demanding than they would be for electrical power production by fusion.

The concept of fast-spectrum, sub-critical nuclear reactors driven by tokamak D-T fusion neutron sources based on ITER physics and technology has been examined. The transuranics-fueled reactors would produce 3000 MWth, which would enable them to fission all the transuranics (TRU) discharged annually from three 1000 MWe Light Water Reactors. The fusion neutron source strength that is required to maintain a constant fission reactor power output as the neutron multiplication constant, k_m, of the sub-critical reactor decreases with fuel burnup

Table 12.5 Tokamak neutron source parameters.

Parameter	SABR nominal	SABR extended	ITER	ARIES-AT
Major Radius R, m	3.75	3.75	6.2	5.2
Current I, MA	8.3	10.0	15.0	13.0
P_{fus}, MW	180	500	400	3000
Neutron Source, 10^{19}/s	7.1	17.6	14.4	106
Aspect Ratio A	3.4	3.4	3.1	4.0
Elongation κ	1.7	1.7	1.8	2.2
Magnetic Field B, T	5.7	5.7	5.3	5.8
B_{TFC}/B_{OH} at coils, T	11.8/13.5	11.8/13.5	11.8/13.5	11.4/
Safety Factor q_{95}	3.0	4.0	3.0	
Confinement H_{IPB98}	1.0	1.06	1.0	2.0
Normalized Beta β_N	2.0	2.85	1.8	5.4
Plasma Energy Mult. Q_p	3	5	5–10	>30
CD eff., $\gamma_{cd}.10^{-20}$ A/Wm^2	0.61	0.58		
Bootstrap cur. frac f_{bs}	0.31	0.26		0.91
Neutron Γ_n, MW/m^2	0.6	1.8	0.5	4.9
FW Heat q_{fw}, MW/m^2	0.23	0.65	0.5	1.2
Availability, %	76	76	25	>90

is given by

$$P_{fusion} \approx \frac{E_{fusion}}{E_{fission}} \nu \cdot \frac{(1-k_m)}{k_m} P_{fission}$$

where E_x is the energy released in the respective x-event and ν is the average number of neutrons released in a fission event.

Nuclear fuel cycle studies indicate that it is practical to achieve >90% transuranics (TRU) burnup with a neutron source strength P_{fus} < 400 MW by repeatedly burning ≈25% of the TRU fuel, then recycling and reprocessing the TRU fuel to remove neutron absorbing fission products and adding fresh TRU. The ≈25% fuel burnup per fuel cycle is limited by an assumed ≈200 dpa radiation damage limit on structural material. If structural materials are developed with >200 dpa radiation damage limit, then the percentage fuel burnup per fuel cycle could be increase by increasing the fusion neutron source strength. A fusion neutron source that produced 400–500 MW would suffice to achieve >90% TRU fuel burnup. The parameters of a tokamak fusion neutron source that would produce 200–500 MW fusion power to meet the requirements of a sub-critical fission transmutation reactor are compared with ITER and ARIES-AT parameters in Table 12.5. The physics and technology of this neutron source are based on the ITER design database. A schematic of the annular sub-critical reactor surrounding the tokamak fusion neutron source is shown in Figure 12.8.

12.7
Future Stellarator Reactors

Although the stellarator confinement concept is not as well-developed as is the tokamak confinement concept, there is a sufficient physics database to project

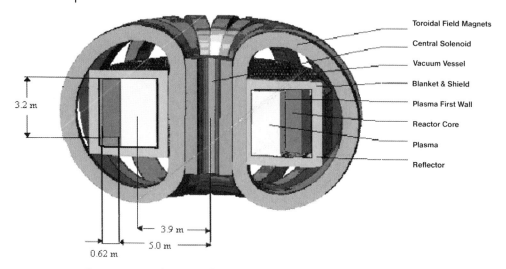

3.2 m

3.9 m
5.0 m
0.62 m

Toroidal Field Magnets
Central Solenoid
Vacuum Vessel
Blanket & Shield
Plasma First Wall
Reactor Core
Plasma
Reflector

Figure 12.8 Configuration of the SABR transmutation of nuclear waste reactor design, with a tokamak fusion neutron source based on ITER physics and technology (Reproduced with permission from the Georgia Tech FRC website.)

the characteristics of a future reactor. Early stellarator reactor studies based on extrapolation from the soon to operate Wendelstein7-X experiments resulted in quite large physical configurations (major radius R $\approx$ 18–22 m), and reactors extrapolated from the somewhat different Large Helical Device are not that much smaller (major radius R $\approx$ 14 m). This unfavorable reactor prospect compared with the more developed tokamak concept (major radius R $\approx$ 5–6 m) led recently to the development of a compact stellarator concept based on a confinement time of 1 s, which requires a factor of $\approx$4 improvement over the widely used ISS95 stellarator confinement scaling correlation.

The magnetic coil system and the resulting plasma shape for a reactor design based on the Wendelstein 7-X machine presently under construction are depicted in Figure 12.9.

HSR Reactor Based on Wendelstein 7-X

- R = 22 m, B_0 = 5 T, B_{max} = 10 T (NbTi SC coils)
- <β> = 5%; based on conservative physics, technology

Figure 12.9 Plasma and coil configuration in a stellarator reactor design (Reproduced with permission from the ARIES website.)

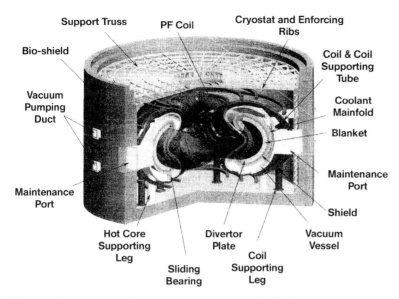

Support Truss   PF Coil   Cryostat and Enforcing Ribs

Bio-shield

Coil & Coil Supporting Tube

Vacuum Pumping Duct

Coolant Mainfold

Blanket

Maintenance Port

Maintenance Port

Shield

Hot Core Supporting Leg   Divertor Plate   Vacuum Vessel

Sliding Bearing   Coil Supporting Leg

Figure 12.10 Compact stellarator reactor design (Reproduced with permission from the ARIES website.)

A more ambitious compact stellarator reactor design is depicted in Figure 12.10 and described by the parameters in Table 12.6.

Four liquid tritium breeder based blanket systems and one solid breeder based system were investigated for the compact stellarator reactor design study: (i) self-cooled flouride-lithium-beryllium (FLIBE) with oxide dispersion strengthened (ODS) ferritic steel structural system; (ii) self-cooled lithium-lead with SiC composite structure: (iii) dual-cooled lithium-lead (or lithium) with helium and ferritic steel structure; and (iv) Li_4SiO_4 with beryllium multiplier and ferritic steel structure.

12.8
Future Spherical Torus Reactor

While the spherical torus confinement concept is not nearly as well-developed as the tokamak (or even the stellarator) confinement concept, recent experiments and

Table 12.6 Compact stellarator NCSX-1 reactor design parameters.

Major radius, R (m)	6.22
Minor radius, a (m)	1.38
B in plasma (T)	6.48
B at coil (T)	12.65
H_ISS95	4.15
n ($10^{20}/m^3$)	3.51
β (%)	6.09

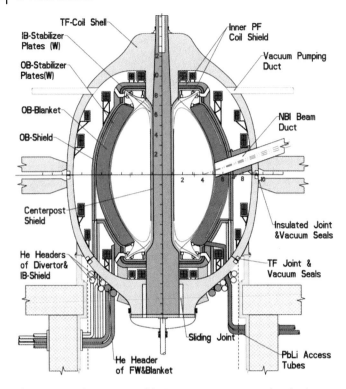

Figure 12.11 Elevation view of the ARIES-ST reactor (Reproduced with permission from the ARIES website.)

theoretical studies have generated an interest in investigating the potential of the concept as a fusion reactor. The design concept for a 1000 MWe spherical torus reactor with a major radius R = 3.2 m is shown in Figure 12.11. This configuration, which is significantly smaller than the ARIES-AT tokamak reactor design, is predicted to be capable of achieving $\beta = 50\%$ (as compared to 6–10% in tokamak reactors). The toroidal field is 7.4 T at the coil and 2.1 T in the plasma. Although the total current is 28 MA, the almost perfect alignment of bootstrap and equilibrium current density profiles results in a requirement of only 28 MW of current-drive power. An advanced "dual-cooled" tritium breeding blanket with flowing lithium-lead breeder and ferritic steel structure cooled with helium is used.

While the technical problems associated with the proximity of the center-post coil to the plasma are considerable, and the physics extrapolations from present experiments are large, the small size and high β projected for this design is attractive.

12.9
Future Reversed Field Pinch Reactor

The reversed field pinch confinement concept is not yet at the stage of development at which its reactor manifestation can be described with confidence. The physics of

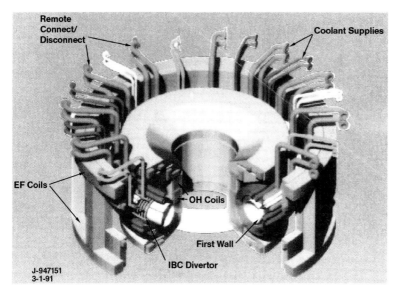

Figure 12.12 TITAN RFP reactor design (Reproduced with permission from the ARIES website.)

confinement scaling, plasma transport and the possibility of eliminating the in-stabilities (with or without a close-fitting conducting shell) which currently limit the discharges to millisecond durations remain to be worked out. Nevertheless, the potential benefit of a compact, high power density magnetic fusion reactor concept has motivated reactor design studies to quantify these benefits and to identify the engineering issues associated with the high power density (e.g., neutron wall loads of 10–20 MW/m^2). The TITAN design is based on the assumption that (with or without a close-fitting conducting shell) a 2–3 order of magnitude increase over the present energy confinement time and pulse duration can be achieved. As opposed to the tokamak and stellarator reactor designs, the TITAN design uses normal conducting magnets, which allows a thinner shield to be used to enhance the compactness. The tritium breeding material is self-cooled lithium, and a vanadium alloy structure is used. The configuration and major parameters are indicated in Figure 12.12. The major parameters are R = 3.9 m, a = 0.6 m, I = 18 MA, β_θ = 23%, τ_E = 0.15s, P_{fus} = 2300 MW.

Problems

12.1. Calculate the DT fusion power produced by a tokamak reactor with R = 5.0 m, a = 1.8 m, κ = 1.8, I = 10 MA, B_φ = 5 T operating at the β-limit.

12.2. Calculate the startup volt-second requirement for the reactor of problem 12.1.

12.3. Calculate flux core radius, r_v, and the thickness, Δ_{OH}, of a central solenoid that could provide the startup volt-seconds of problem 12.2 with a maximum field of B_{OH} = 12 T while satisfying the ASME Code requirements for S_m = 200 MPa.

12.4. Calculate the field at the TF coil that is required to produce $B_\varphi = 5\,T$ at the plasma centerline ($R = 5\,m$). Assume that the radius to the wall of the plasma chamber $r_w = 1.1a$ and that the inboard blanket plus shield thickness $\Delta_{BS} = 1.1\,m$.

12.5. Calculate the peak heat flux to the first wail and to the divertor plate if the fraction of the plasma exhaust power going to the divertor is $f_{div} = 0.5$, the fraction of this power entering the divertor that reaches the plate is $f_{dp} = 0.5$, the divertor plate and first wall peaking factors are $f_{dp}^* = 10$ and $f_{FW}^* = 1.3$, the fraction of the area surrounding the plasma that is subtended by the divertor is $\varepsilon_{div} = 0.1$, the width of the divertor plate is $\Delta_{div} = 10\,cm$ and the auxiliary heating power is 50 MW.

12.6. Calculate the thickness range of the first wall tube that can satisfy the ASME stress limits of Eqs. (19–27) and (19–29) on a "tube bank" first wall for the peak heat flux of problem 12.5. Consider two cases: a) stainless steel ($S_m = 110\,MPa$, $M = \alpha E / \kappa (1 - \nu) = 0.22\,MPa\text{-}m/W$) first wall tubes of radius $r_c = 2.5\,cm$ and water coolant at 10 MPa pressure; and b) vanadium alloy ($S_m = 146\,MPa$, $M = \alpha E / \kappa (1 - \nu) = 0.055\,MPa\text{-}m/W$) first wall tubes of radius $r_c = 2.5\,cm$ with lithium coolant at 1 MPa pressure. (Neglect the volumetric neutron heating q'''.)

12.7. Calculate the power flux of 14 MeV DT fusion neutrons incident on the first wall of the reactor of problem 12.1.

Appendix A: Frequently Used Physical Constants

c	Speed of light in vacuum	$2.99\,793 \times 10^8$ m/sec
ε_0	Permittivity of vacuum	8.854×10^{-12} F/m
μ_0	Permeability of vacuum	1.257×10^{-6} H/m
a	Stefan-Boltzmann constant	5.67×10^{-8} J/m^2 sec K^4
h	Planck constant	6.625×10^{-34} J sec
k	Boltzmann constant	1.3804×10^{-23} J/K
A	Avogadro constant	6.025×10^{23}/mol (physical)
e	Charge of proton	1.6021×10^{-19} C
1 ev	Electronvolt	1.6021×10^{-19} J
eV/k		11,600 K

Particle	Mass ($10^{-27} \times$ kg)	Rest Energy (MeV)
e	9.108×10^{-4}	0.511
n	1.674	939.512
H^1 atom	1.673	938.730
H^2 atom	3.343	1876.017
H^3 atom	5.006	2809.272
He^3 atom	5.006	2809.250
He^4 atom	6.643	3728.189
Li^6 atom	9.984	5602.735
Li^7 atom	11.64	6534.995

Fusion: Second, Completely Revised and Enlarged edition. Weston M. Stacey
Copyright © 2010 Wiley-VCH Verlag GmbH & Co. KGaA, Weinheim
ISBN: 978-3-527-40967-9

Appendix B: Energy Conversion Factors

Energy Conversion Factors

	erg	joule = w-sec	kilowatt-hour	cal	Btu	kg-m	electronvolt	1 mol electronvolt	10^{-3} unit atomic mass	g-mass equivalent	megawatt-day
1 erg	1	1.000×10^{-7}	2.78×10^{-14}	2.39×10^{-8}	9.48×10^{-11}	1.02×10^{-8}	6.24×10^{11}	1.04×10^{12}	6.70×10^{5}	1.11×10^{-21}	1.16×10^{-18}
1 joule	1.00×10^{7}	1	2.78×10^{-7}	2.39×10^{-1}	9.48×10^{-4}	1.02×10^{-1}	6.24×10^{18}	1.04×10^{-5}	6.70×10^{12}	1.11×10^{-14}	1.16×10^{-14}
1 kilowatt-hour	3.60×10^{13}	3.60×10^{6}	1	8.59×10^{5}	3.41×10^{3}	3.67×10^{5}	2.25×10^{25}	3.73×10^{1}	2.41×10^{19}	4.01×10^{-8}	4.17×10^{-5}
1 cal	4.19×10^{7}	4.19	1.16×10^{-6}	1	3.97×10^{-3}	4.27×10^{-1}	2.62×10^{19}	4.34×10^{-5}	2.81×10^{13}	4.66×10^{-14}	4.85×10^{-11}
1 Btu	1.05×10^{10}	1.05×10^{3}	2.93×10^{-4}	2.52×10^{2}	1	1.08×10^{2}	6.58×10^{21}	1.09×10^{-2}	7.07×10^{15}	1.17×10^{-11}	1.22×10^{-8}
1 kilogram-meter	9.81×10^{7}	9.81	2.72×10^{-6}	2.34	9.30×10^{-3}	1	6.12×10^{19}	1.02×10^{-4}	6.57×10^{13}	1.09×10^{-13}	1.14×10^{-10}
1 Electronvolt	1.60×10^{-12}	1.60×10^{-19}	4.45×10^{-26}	3.82×10^{-20}	1.52×10^{-22}	1.63×10^{-10}	1	1.66×10^{-24}	1.07×10^{-6}	1.78×10^{-33}	1.85×10^{-30}
1 mol electronvolt	9.65×10^{-11}	9.75×10^{-4}	2.68×10^{-2}	2.30×10^{4}	9.15×10^{1}	9.84×10^{3}	6.02×10^{23}	1	6.47×10^{17}	1.07×10^{-9}	1.12×10^{-6}
10^{-3} unit atomic mass	1.49×10^{-6}	1.49×10^{-13}	4.15×10^{-20}	3.56×10^{-14}	1.42×10^{-16}	1.52×10^{-14}	9.31×10^{5}	1.55×10^{-18}	1	1.66×10^{-27}	1.73×10^{-24}
1 g-mass equivalent	8.99×10^{20}	8.99×10^{13}	2.50×10^{7}	2.15×10^{13}	8.52×10^{10}	9.16×10^{12}	5.61×10^{32}	9.32×10^{8}	6.02×10^{23}	1	1.04×10^{3}
1 Megawatt-day	8.64×10^{17}	8.64×10^{10}	2.40×10^{4}	2.06×10^{10}	8.19×10^{9}	8.81×10^{9}	5.39×10^{29}	8.95×10^{5}	5.79×10^{23}	9.61×10^{-4}	1

Fusion: Second, Completely Revised and Enlarged edition. Weston M. Stacey
Copyright © 2010 Wiley-VCH Verlag GmbH & Co. KGaA, Weinheim
ISBN: 978-3-527-40967-9

Appendix C: Engineering Conversion Factors

Unit Conversion Factors

Temperature

$$1\,^\circ C = 1.8\,^\circ F \qquad\qquad 1\,^\circ F = 0.556\,^\circ C$$

$$n\,^\circ C = \left(\tfrac{9}{5}n + 32\right)\,^\circ F \qquad n\,^\circ F = \left[(n-32)\tfrac{5}{9}\right]\,^\circ C$$

Pressure

$$1\,kg/cm^2 = 14.223\,psi \qquad 1\,mm\ Hg\ at\ 0\,^\circ C = 1.361\,cm\ H_2O\ at\ 15\,^\circ C$$

$$1\,psi = 0.07\,031\,kg/cm^2 \quad 1\,cm\ H_2O\ at\ 15\,^\circ C = 0.7349\,mm\ Hg\ at\ 0\,^\circ C$$

Flow

$$1\,kg/sec = 3.60\,t/hr = 7936.6\,lb/hr$$

$$1\,t/hr = 0.2778\,kg/sec = 2204.6\,lb/hr$$

$$1\,lb/hr = 1.260 \times 10^{-4}\,kg/sec = 4.536 \times 10^{-4}\,t/hr$$

$$1\,L/sec = 15.850\,(US)gal/min = 13.199\,Imp\,gal/min$$

$$1\,(US)\,gal/min = 0.8327\,Imp\,gal/min = 0.063\,091\,L/sec$$

$$1\,Imp\,gal/min = 1.20\,094\,(US)\,gal/min = 0.075\,766\,L/sec$$

Heat Energy

$$1\,kcal = 3.9683\,Btu \qquad 1\,Btu = 0.2520\,kcal$$

Heat Flux

$$1\,cal/cm^2\ sec = 4.1840\,W/cm^2 = 13\,263\,Btu/ft^2\ hr$$

$$1\,W/cm^2 = 0.2390\,cal/cm^2\ sec = 3171\,Btu/ft^2\ hr$$

$$1\,Btu/ft^2\ hr = 0.754 \times 10^{-4}\,cal/cm^2\ sec = 3.154 \times 10^{-4}\,W/cm^2$$

Fusion: Second, Completely Revised and Enlarged edition. Weston M. Stacey
Copyright © 2010 Wiley-VCH Verlag GmbH & Co. KGaA, Weinheim
ISBN: 978-3-527-40967-9

Heat Transfer Coefficient

$$\text{cal/cm}^2 \text{ sec } ^{\circ}\text{C} = 4.1833 \text{ W/cm}^2 \, ^{\circ}\text{C} = 7373.5 \text{ Btu/ft}^2 \text{ hr } ^{\circ}\text{F}$$

$$1 \text{ W/cm}^2 \, ^{\circ}\text{C} = 0.239 \text{ cal/cm}^2 \text{ sec } ^{\circ}\text{C} = 1762.2 \text{ Btu/ft}^2 \text{ hr } ^{\circ}\text{F}$$

$$1 \text{ Btu/ft}^2 \text{ hr } ^{\circ}\text{F} = 1.3563 \times 10^{-4} \text{ cal/cm}^2 \text{ sec } ^{\circ}\text{C} = 5.673 \times 10^{-4} \text{ W/cm}^2 \, ^{\circ}\text{C}$$

Length

$1 \text{ cm} = 0.3937 \text{ in.}$	$1 \text{ in.} = 2.540 \text{ cm}$
$1 \text{ m} = 3.28\ 083 \text{ ft}$	$1 \text{ ft} = 0.30\ 480 \text{ m}$
$1 \text{ km} = 0.62\ 137 \text{ mile}$	$1 \text{ mile} = 1.609\ 3 \text{ km}$

Area

$1 \text{ cm}^2 = 0.1550 \text{ in.}^2$	$1 \text{ in.}^2 = 6.4516 \text{ cm}^2$
$1 \text{ m}^2 = 10.764 \text{ ft}^2$	$1 \text{ ft}^2 = 0.09\ 290 \text{ m}^2$

Volume

$1 \text{ cm}^3 = 0.06\ 103 \text{ in.}^3$	$1 \text{ in.}^3 = 16.3870 \text{ cm}^3$
$1 \text{ m}^3 = 35.314 \text{ ft}^3$	$1 \text{ ft}^3 = 0.028\ 317 \text{ m}^3$

Capacity

$$1 \text{ L} = 0.21\ 998 \text{ Imp gal} = 0.26\ 417 \text{ (US) gal}$$

$$1 \text{ Imp gal} = 4.5460 \text{ L} = 1.20\ 094 \text{ (US) gal}$$

$$1 \text{ (US) gal} = 3.7854 \text{ L} = 0.8327 \text{ Imp gal}$$

Weight

$1 \text{ kg} = 2.2046 \text{ lb}$	$1 \text{ lb} = 0.4536 \text{ kg}$

Density

$1 \text{ g/cm}^3 = 62.43 \text{ lb/ft}^3$	$1 \text{ lb/ft}^3 = 0.01\ 602 \text{ g/cm}^3$

Power

$1 \text{ hp} = 0.7457 \text{ kW}$	$1 \text{ kW} = 1.341 \text{ hp}$

Index

Fusion: Second, Completely Revised and Enlarged edition. Weston M. Stacey
Copyright © 2010 WILEY-VCH Verlag GmbH & Co. KGaA, Weinheim
ISBN: 978-3-527-40967-9